Geologic Cross Sections

CYNTHIA SHAUER LANGSTAFF
DAVID MORRILL

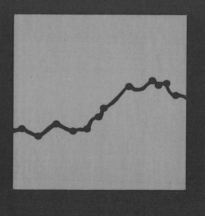

Geologic Cross Sections

Cynthia Shauer Langstaff
David Morrill

International
Human
Resources
Development
Corporation
Boston

Acknowledgments

The authors of this book wish to thank the following individuals for their advice, assistance, and contribution:

Howard B. Bradley, Mobil Exploration and Producing Services Inc., Dallas and *Derek F. Harvey*, Mobil Oil Corporation, New York for their constant support and guidance throughout the project;

D. Leroy Sims, W. Thomas Deubel, Joseph C. DuBois, and *W.I. Mushake*, of Mobil Exploration and Producing Services Inc., Dallas and *Alan J. Witherspoon*, Mobil Oil Corporation, Denver for their advice on content and review of the manuscript;

Prof. Marland P. Billings, Department of Geological Sciences, Harvard University, for recommendations on appropriate terminology;

Michael Hays, Gail Smith, Carolyn Yoder, and *Laura Grunwerg* of IHRDC Publications for their invaluable review, editorial and graphic assistance and especially to *Mary Rahmati* of IHRDC for her illustrations and for her total dedication to the project.

Copyright ©1981 by International Human Resources Development Corporation. All rights reserved. No part of this book may be used or reproduced in any manner whatsoever without written permission of the publisher except in the case of brief quotations embodied in critical articles and reviews. For information address: IHRDC, Publishers, 137 Newbury Street, Boston, MA 02116.

ISBN: 0-934634-22-X

Printed in the United States of America

Contents

1 Introduction — 1

2 The Elements of a Cross Section — 6
    2.1 Data — 7
    2.2 Line of Section — 12
    2.3 Scale — 20
    2.4 Datum — 27

3 The Construction of 2-Dimensional Cross Sections — 31
    3.1 Structural Cross Sections — 33
    3.2 Stratigraphic Cross Sections — 38
    3.3 Correlation and Geological Interpretation — 40
    3.4 Completing the Cross Section — 44

4 3-Dimensional Diagrams — 47
    4.1 Fence Diagrams — 48
    4.2 Block Diagrams — 52

5 Computer-Drawn Cross Sections — 56

6 Review — 71

7 Supplemental References — 74

8 Questions and Exercises — 76

9 Solutions — 96

Preface

This book accompanies a videotape program of the same name. The combined videotape and book, referred to as a module of instruction, was one of three prepared by IHRDC on a joint basis with Mobil Oil Corporation during 1980. The three modules, one each in geology, geophysics and petroleum engineering, were produced to determine whether this medium of instruction would provide an effective way of teaching recent graduates and those individuals changing specialties, "what they need to know, when they need to know it." The major observations of the pilot production stage were that properly designed and properly used, video-assisted instruction is effective, efficient, and convenient.

With the confidence that this instructional medium provides one way for the international petroleum industry to train young graduates in exploration and production, IHRDC sought financial and advisory support from a limited number of companies to undertake the development of the *Basic Technical Video Library for the E&P Specialist*. To date the following companies have agreed to serve as Sponsors: Mobil, AGIP, ARAMCO, Cities Services, Dome Petroleum Ltd., Gulf, Phillips, Standard Oil of California/ Chevron, and Texaco.

Work on the Library began in July 1981. With an accelerated production schedule of 24 modules per year and the continued support of the Sponsors, the Library should be completed in about five years. Participation in the development of the Library is open to other Sponsors.

1
Introduction

A cross section is a profile showing geological features in a vertical plane through the earth. Some geologists prefer the term "section" for this type of diagram, reserving "cross section" to denote a section made perpendicular to structural strike. In practice, relatively few of these diagrams can be constructed strictly perpendicular to strike. Also, the word "section" used alone could be misconstrued to mean seismic section, columnar section, or even thin section. In petrology, a section is a lithologic sequence which can be viewed in outcrop. In paleontology, a section is an important division of a genus. On public lands, a section is a smaller division of a township. To avoid ambiguity, many geologists have come to use the term "cross section" to describe a geological profile made across the earth, i.e., along any vertical plane through the earth, regardless of its orientation to structural strike. The term "dip section" is used to indicate cross sections made perpendicular to strike. We have adopted this more general definition of "cross section" since we believe that any practical study guide should reflect the most common and current usage of terms.

There are two categories of cross sections: structural and stratigraphic. Structural cross sections illustrate present-day structural features such as dips, folds, and faults. Stratigraphic cross sections show characteristics such as formation thicknesses, lithologic sequences, stratigraphic correlations, facies changes, unconformities, fossil zones, and ages. Simplification of structural elements permits greater emphasis on stratigraphic relationships in a stratigraphic cross section.

Since cross sections are drawn to scale, the locations of features on the diagram are measured off with respect to an arbitrary horizontal reference line called a datum. Structural and stratigraphic cross sections differ fundamentally in the type of datum line used. Structural cross sections are scaled off from a datum which represents an elevation; stratigraphic cross sections are based on a datum which represents a flattened stratigraphic boundary. We will discuss datum lines in more detail in chapter two.

Cross sections are useful in the search for potential hydrocarbon traps which exist at depth in the earth. The petroleum explorationist uses cross sections both as working diagrams for problem-solving and as finished illustrations for display. As working diagrams, cross sections are helpful in the visualization of regional and local geological relationships. Simplification to two dimensions often facilitates interpretation, especially in geologically complex areas. Interpretive problems are frequently resolved during the working and reworking of one or more cross sections.

Cross sections are also useful for display purposes. Diagrammatic cross sections show broad relationships and serve to orient the

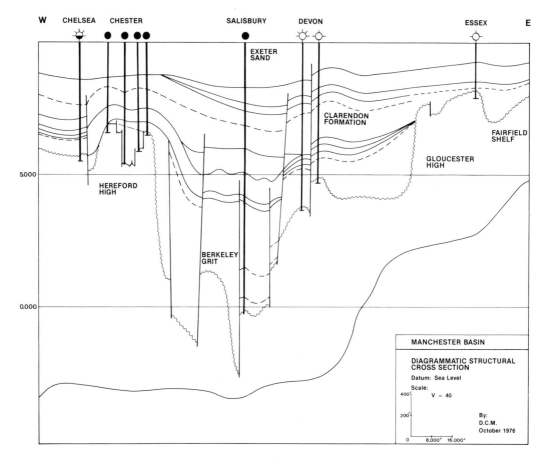

FIGURE 1.1
Diagrammatic structural cross section.

audience to the general geology of a region (figure 1.1). More detailed cross sections usually cover a smaller area and are invaluable in communicating the geologist's interpretation of a prospect (figures 3.1 and 3.6). Such cross sections complement geological maps and geophysical data in an integrated exploration program.

2
The Elements of a Cross Section

The geologist must answer a number of fundamental questions and must make a number of important decisions before a cross section is constructed. What will the cross section illustrate? Will it be a structural or a stratigraphic cross section? Will it be used to communicate broad relationships or to illustrate great detail? Is it for problem-solving or for display? What are the geological features of interest? What data will be used? How should the scale be chosen? Each decision should be made with the ultimate purpose of the diagram in mind. Such planning is the key to an effective cross section.

2.1 Data

Different types of data can be used in developing a cross section. The diagram may be based on outcrop information, or on subsurface data derived from wells and geophysical surveys, or it may use both outcrop and subsurface data.

Cross sections which use outcrop information are particularly useful in frontier areas of petroleum exploration, where subsurface information is often sparse or unavailable. In covered areas between outcrops, the geologist can infer information from soil type, vegetation, and topography. Shallow data control is therefore fairly continuous wherever surface geology is used.

Many geological cross sections made for petroleum exploration rely on subsurface data. The information for such cross sections comes primarily from wells, and thus the data is not continuous over the length of the cross section. Supplemental geophysical data can be helpful in interpreting between the data points. It is generally desirable to have data points relatively evenly spaced along the line of section, wherever this is permitted by well control. However, a greater density of information may be necessary over a feature of special interest. For instance, additional data points are often chosen over a fold or on each side of a known or suspected fault. The reliability of data is another important factor which must be considered. All data should be checked for accuracy. Incorrect well location or elevation information is a common source of error, especially in old wells. This information is easily verified in areas of relief by plotting

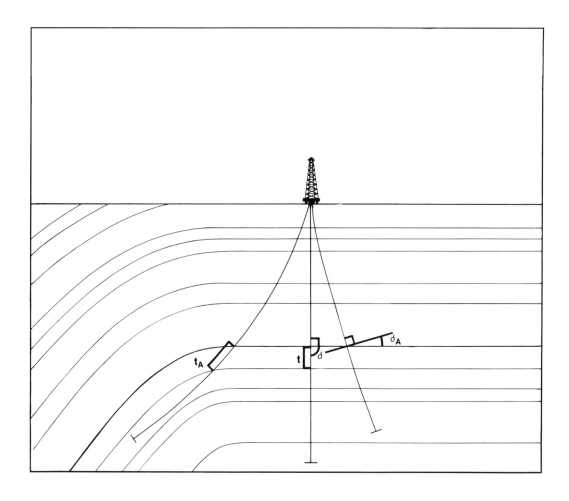

FIGURE 2.1
Distortion of true thickness (t) and true dip (δ) measurements due to borehole deviation; t_A = apparent thickness and $δ_A$ = apparent dip.

the wells on topographic maps to determine if the given elevation is reasonable.

Incorrect stratigraphic tops are another possible source of error. Drillers' picks are particularly prone to mistakes and should be compared with adjacent wells. Wherever possible, log tops should be used in place of drillers' picks. The geologist should recorrelate all logs to ensure accuracy and consistency. A third source of error is borehole deviation, i.e., when the well is not drilled vertically, but at some angle to the vertical (figure 2.1). Depth and dip measurements assume that the hole was vertical and are therefore distorted by borehole deviation. The distortion can be corrected only if a downhole directional survey exists. Deviation in recent wells is usually intentional and directional surveys are usually available. If such a survey exists, true depth and dip can be calculated and incorporated into the cross section. If the deviated well happens to lie in or near the plane of section, the actual trace of the hole can be drawn on the section with the help of the directional survey. If the borehole does not lie in the plane of section, or if it is deviated in three dimensions, it must be projected onto the cross section. This projection of the borehole will show apparent, rather than actual deviation (figure 2.2). One should always be cautious in accepting the conclusions of previous studies. Many authors from the academic community who write "definitive" papers have access to less data than do petroleum industry personnel. This is not to say that industry-generated studies should be blindly preferred to academic studies. The geologist should screen every potential resource carefully. The expertise of authors should be

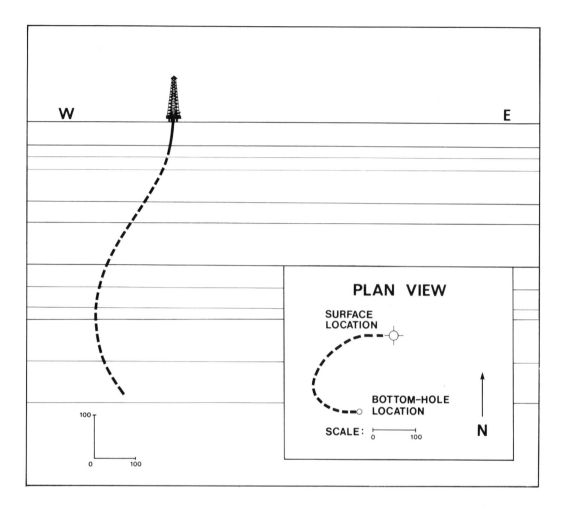

FIGURE 2.2
Cross-sectional projection of a borehole deviated in three dimensions. Segments of the borehole which do not lie in the plane of section are dashed onto the cross section. Due to foreshortening, the dashed segments show only apparent deviation. This can be seen by comparison with the inset plan view.

assessed; the extent and reliability of their data should be checked; and the validity of their conclusions should be scrutinized. Occasionally, it is useful to verify maps and cross sections with small sketches. Seismic interpretations should be reviewed, usually with the help of a geophysicist. Even in-house studies should be subject to this examination. Careful control over the quality of data input is essential to the construction of an accurate cross section.

2.2 Line of Section

The next step is to choose the line of section, i.e., the line along which the cross section would intersect the surface of the earth (figure 2.3). The approximate location of the line of section is largely determined by the location of the features of interest. The orientation of the line is

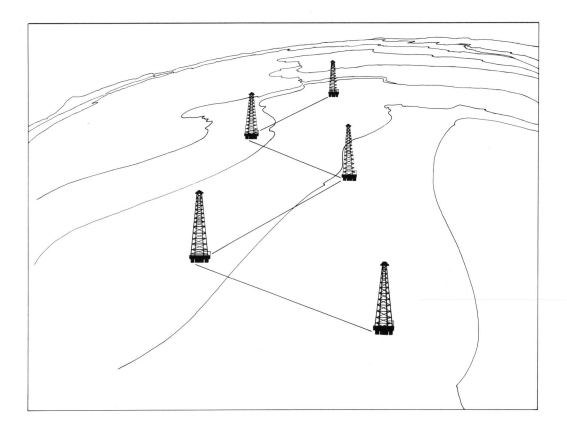

FIGURE 2.3
Line of section.

primarily controlled by the nature of the geological relationships to be illustrated and by the availability of reliable data.

Cross sections may trend perpendicular to strike, oblique to strike, or parallel to strike (figure 2.4). These cross sections are termed dip sections, oblique sections, and strike sections, respectively. One must bear in mind that the dip shown on an oblique or strike section is always less than true dip. The greater the angle between the true dip direction and the line of section (β), the greater the difference between the true dip (δ) and apparent dip (δ_A). Mathematically, the relationship between these variables is expressed as:

(2.1) $\tan\delta_A = \tan\delta \cos\beta$.

Minor manipulation of equation 2.1 will demonstrate that apparent dip may vary between horizontal, where the line of section is parallel to strike ($\beta = 70°$), and true dip (δ), where the line of section is perpendicular to strike ($\beta = 0$) (figure 2.5). Thus, quantitative dip measurements will be distorted in cross sections which are not made perpendicular to strike.

Since it is unusual to find a succession of wells along a straight line, many cross sections which use well data follow zig-zag paths (figure

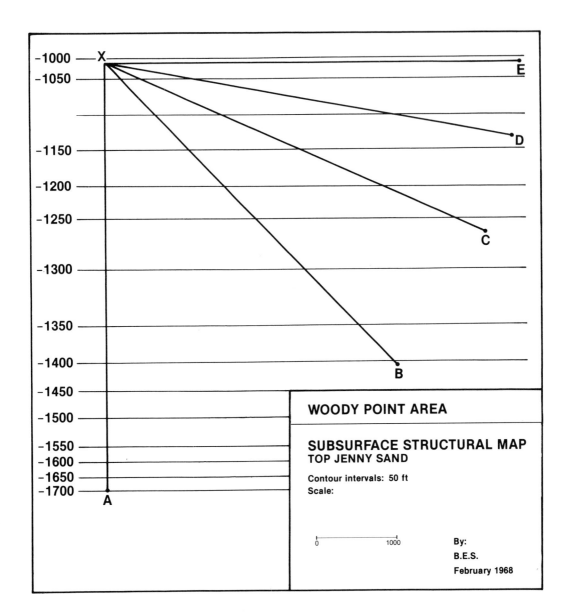

FIGURE 2.4
Lines of section with different orientations to structural strike: perpendicular (XA), oblique (XB, XC, XD), and parallel (XE).

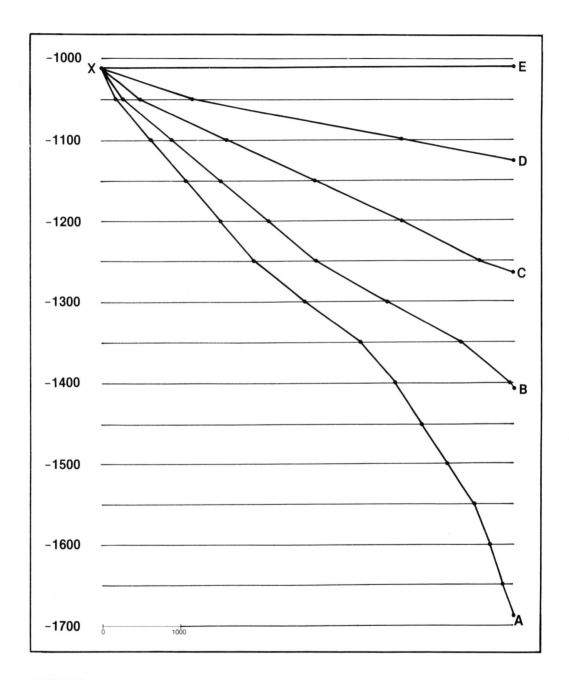

FIGURE 2.5
Vertical profiles of the top of the jenny sand, made along the lines of section illustrated in figure 2.4. True dip is shown by profile XA which trends perpendicular to structural strike.

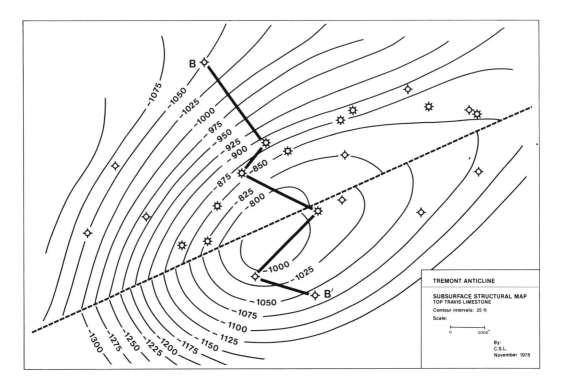

FIGURE 2.6
Zig-zag line of section.

2.6). Lines of section may change direction or even trend in a circle if necessary to illustrate the desired geological features (figure 2.7).

The apparent dip of horizons on a cross section will vary from segment to segment as the orientation of the line of section changes. Thus, an irregular line of section will affect the shape of the geological features illustrated in a cross section (figure 2.8). It is desirable to choose as straight a line of section as possible.

Some geologists project nearby wells into a straight line of section instead of deflecting the line of section to the wells. It is usually necessary to construct a subsurface

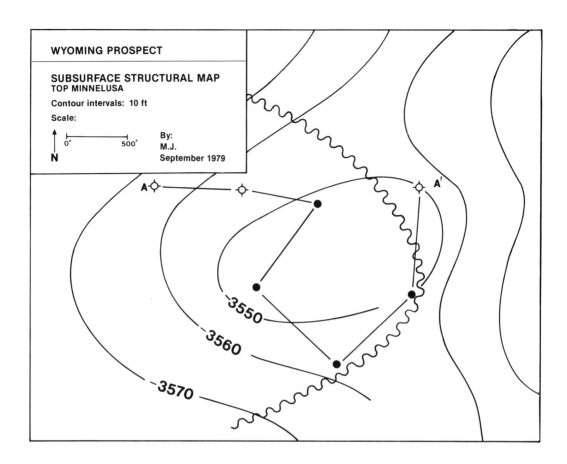

FIGURE 2.7
Line of section across a field in Wyoming. The circular path was necessary in order to illustrate the nature of the stratigraphic hydrocarbon trap.

contour map before projecting wells. The location of the well on the cross section is determined by a normal from the well to the line of section, or by projecting the well along strike (figure 2.9). The well is dashed onto the cross section to distinguish it from any wells which actually lie in the plane of section (figure 2.10). A correction is made for any change in the apparent dip of beds, using equation 2.1. This procedure is useful for diagrammatic structural cross sections, or when projected wells are close to the straight line of section. In

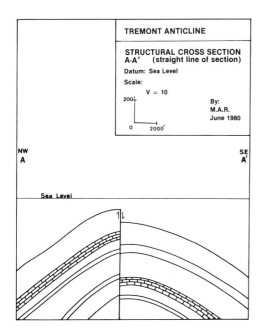 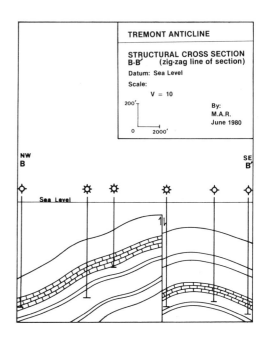

FIGURE 2.8
Two structural cross sections through the same geological feature. Cross section A-A' was made along a straight line of section while cross section B-B' was made along the zig-zag line of section seen in figure 2.6. Note that the shape of the geological feature becomes distorted in B-B' as the orientation of the line of section changes.

other circumstances, it is best to take the line of section right up to the well.

Geology sometimes makes a zig-zag line of section undesirable. If the required well is too far away to project onto the section, information from the well still may be used in interpretation. In such a case, a small cross section to the unillustrated well may be inset above the point to which the well would have been projected. This inset clarifies the geologist's interpretative reasoning.

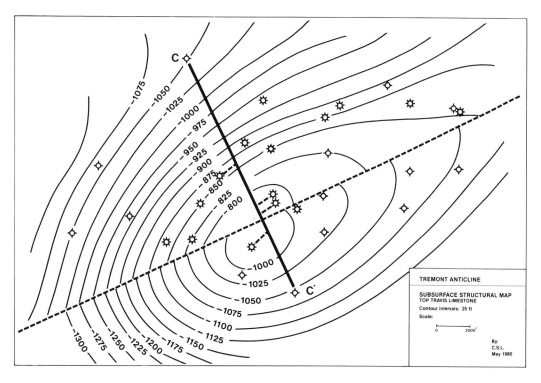

FIGURE 2.9
Wells projected along strike into a straight line of section.

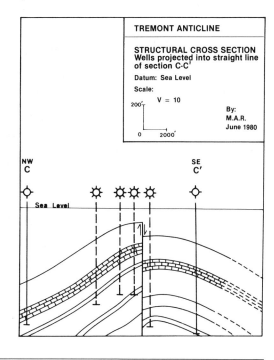

FIGURE 2.10
Structural cross section along the line of section shown in figure 2.9. Projected wells are dashed onto the diagram. Compare with the cross sections in figure 2.8 which illustrate the same faulted anticline.

THE ELEMENTS OF A CROSS SECTION

2.3 Scale

After the line of section and data points have been chosen, the geologist must decide on the scale of the cross section. There are actually two scales to consider: one horizontal and one vertical. For convenience in construction and cross reference, the horizontal scale is often taken to be equal to the scale of corresponding geological maps. The vertical scale may or may not be equal to the horizontal scale, depending on what the cross section is intended to illustrate.

Greater stratigraphic detail can be shown if the vertical scale is large. However, if the horizontal scale were to be made equally large, the cross section might become absurdly long. In such cases, the horizontal scale is taken to be smaller than the vertical scale, a technique called vertical exaggeration.

When using vertical exaggeration, the vertical scale is usually chosen to be some easy multiple of the horizontal scale. The degree of exaggeration in the vertical direction, V, can be expressed as:

(2.2) $$V = \frac{l_v}{l_h},$$

where l_v and l_h are the lengths of a unit distance on the vertical and horizontal scales, respectively.

In petroleum exploration, some vertical exaggeration is usually employed in order to show subsurface features in sufficient detail. However, distortion is introduced when the horizontal and vertical scales are different.

Dips in an exaggerated cross section show a nonlinear deviation from true dip. The relationship between true dip, δ, and exaggerated dip, δ_E, is expressed as:

(2.3) $$\tan\delta_E = V\tan\delta,$$

where V is vertical exaggeration as defined by equation 2.2. Figure 2.11 illustrates this relationship between true and exaggerated dips for different values of V. True dip occurs where $V = 1$ and is thus read off of the bottom edge of the graph.

Note in figure 2.11 that the difference between two very gentle true dips is accentuated by vertical exaggeration, while the difference between two steep true dips is diminished. Vertical exaggeration, therefore, highlights relationships between features which are nearly flat-lying, while it makes it more difficult to distinguish between features with steeper dips. For

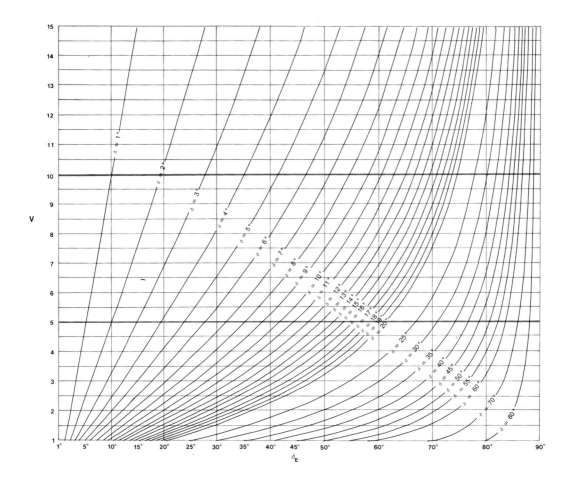

FIGURE 2.11
Solutions to equation 2.3 for different values of V (vertical exaggeration) and δ (true dip). True dip occurs where V = 1. To solve for δ_E (exaggerated dip), locate the solution curve which originates at the given value of δ along the lower edge of the graph. Follow the curve upwards until it intersects with the known value of V. The horizontal component of the point of intersection is δ_E; read the value of δ_E off the horizontal scale at that point.

instance, on a cross section with a vertical exaggeration of 8, beds dipping 3°E and an unconformity with a dip of 6°E will have apparent dips of 23° and 40°, respectively. On the same cross section, a thrust fault with true dip of 39°W and a normal fault with true dip of 63°W will show respective apparent dips of 81° and 86° (figures 2.12 and 2.13).

Formation thicknesses are also distorted by vertical exaggeration (figure 2.14). In an exaggerated cross section, the vertical component of all

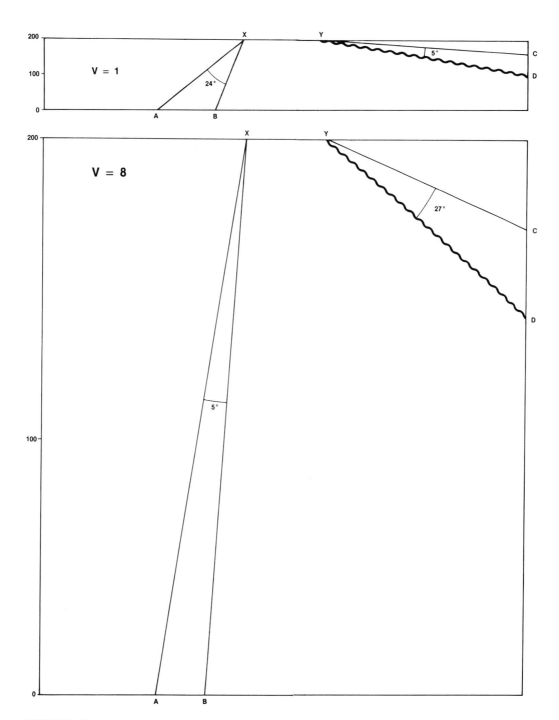

FIGURE 2.12
Effect of vertical exaggeration on the angle between dipping linear features: XA and XB (thrust fault and normal fault); and YC and YD (bedding planes and unconformity). The angle between gently dipping lines is augmented, while the angle between more steeply dipping lines is diminished.

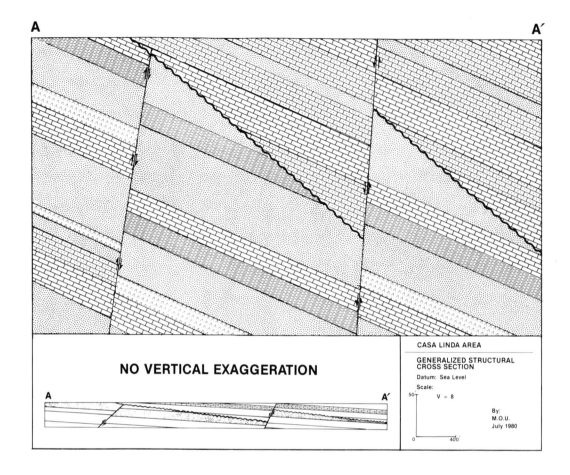

FIGURE 2.13
Effect of vertical exaggeration on dipping features in a cross section. The angle between gently dipping features (bedding planes and an unconformity) is enhanced by vertical exaggeration, while a thrust fault with true dip of 39°W and a normal fault with true dip of 63°W become nearly indistinguishable.

distances is multiplied by the factor V as defined in equation 2.2. When defined in this manner, the factor V does not affect the horizontal component of distances on the cross section. Thus, in figure 2.14, distances measured horizontally remain the same in both the exaggerated (b) and unexaggerated (a) views, while distances measured vertically are multiplied by a factor of V = 2 in the exaggerated view. The exaggerated thickness of strata t_E, will vary as the exaggerated dip of beds changes from horizontal to vertical. When the

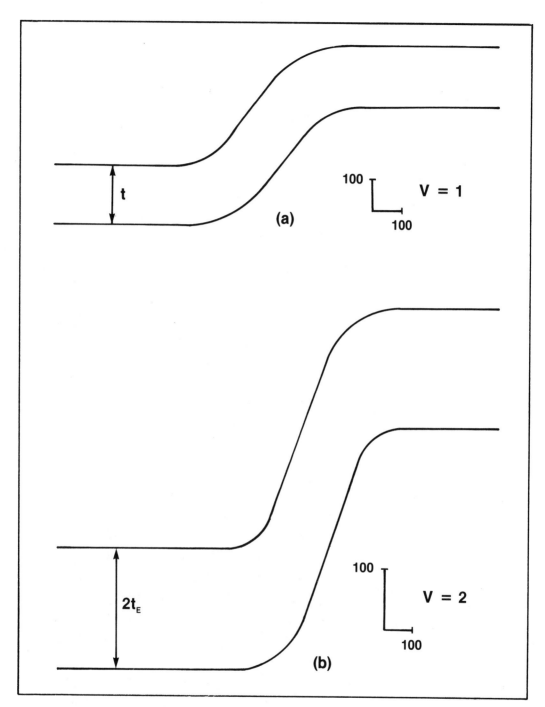

FIGURE 2.14
Apparent attenuation due to vertical exaggeration. Attenuation is greatest where true dip is steepest.

beds are vertical, thickness is measured horizontally and $t_E = t$, where t is true thickness. When the beds are horizontal, thickness is measured vertically and $t_E = Vt$. Thus, in a vertically exaggerated cross section where $V>1$, a bed will appear thickest in areas of gentle dip and will thin as dip increases. This results in an apparent attenuation, or thinning of strata in areas of steepening dip, for instance on the flanks of a fold (figure 2.14).

Of course, vertical exaggeration involves two variables: the vertical and horizontal scales. We could look on the distortion as a shortening in the horizontal direction, in which case the horizontal component of all distances would be modified by a factor of $1/V$. This is merely another way of looking at the same relationship. Whichever point of view you may wish to adopt, the salient point to remember is that vertical exaggeration results in an apparent attenuation or thinning of strata over areas of steeper dip. Quantitative features such as dip and formation thickness are preserved where the horizontal and vertical scales are equal. Although these features are distorted when the scales are not equal, important geological relationships can be emphasized. Used wisely, vertical exaggeration is an effective tool in the construction of cross sections. The decision to employ vertical exaggeration is based on what the cross section is intended to illustrate and on how one can most effectively achieve that illustration. An exaggerated cross section should be clearly labeled with bar scales and the value of V. For quick reference, it may also be helpful to include a small inset with an unexaggerated view of the cross section (figure 2.13).

Practical considerations also affect the choice of scale. The most important factor is that the overall size must be adequate to show the desired detail. In addition, the cross section should be large enough to be visible from a distance if it is intended for display. However, one must take care that the figure will fit available reproducing machines. Overly large cross sections are unwieldy as well as being expensive or even impossible to reproduce.

2.4
Datum

The chosen vertical scale will be used to measure off tops and bases of geological boundaries on the cross section. However, the measurements must be made with respect to some reference line. The line is a one-dimensional representation of a feature or surface which is continuous over the area covered by the cross section. Regardless of the actual relief of the feature or surface, it will be represented by a straight horizontal line, and it will be used as a reference when scaling off boundaries on the cross section. Such a reference line is called a datum line, and the feature or surface represented by the datum line is called a datum plane.

An elevation is one common type of datum plane (figure 2.15). Geological boundaries may be scaled off from the corresponding datum line according to their known elevations in each well. Since elevation planes are already horizontal, existing spatial relationships are preserved and the resulting cross section shows present-day geological structure. Cross sections using an elevation datum are called structural cross sections.

The other type of datum plane is a stratigraphic boundary (figure 2.16). The boundary may be on a marker horizon or on an unconformity. The most important criteria for choosing a stratigraphic datum are that the boundary be easily recognized and reasonably widespread.

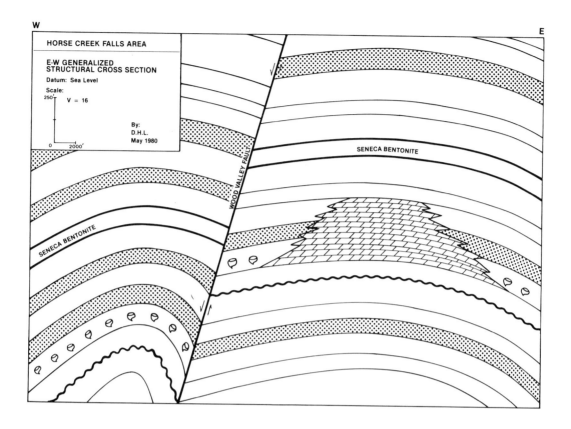

FIGURE 2.15
Generalized structural cross section. Sea level was used as a datum for this cross section.

Few stratigraphic boundaries are horizontal, yet a stratigraphic datum plane is shown as a straight horizontal line on the cross section. For this reason, it is known as a flattened stratigraphic datum. Cross sections which are based on this type of datum are called stratigraphic cross sections. Since the reference boundary is not an elevation, it is no longer possible to simply scale off elevations relative to sea level. Instead, the interval from the datum boundary to the desired horizon is scaled off in each well.

Flattening a stratigraphic boundary causes changes in the

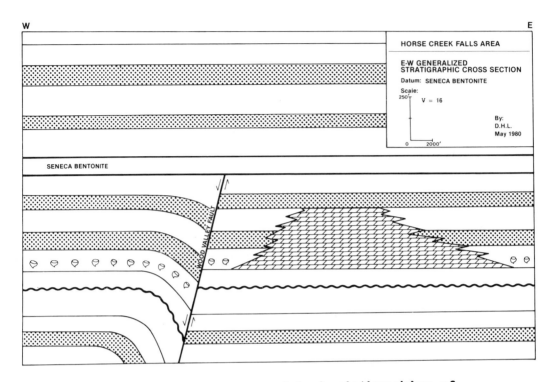

FIGURE 2.16
Generalized stratigraphic cross section. This cross section was developed along the same line of section as the structural cross section in figure 2.15, using the Seneca Bentonite as a stratigraphic datum. The structural elements which can be seen in this figure are paleostructural features which existed at the time of deposition of the Seneca Bentonite.

geometrical relationships of geological features (compare figures 2.15 and 2.16). The resulting cross section illustrates paleostructure, i.e., structural relationships which existed at the time of deposition or formation of the stratigraphic datum horizon. If a cross section is intended for paleostructural analysis, then it is important to choose a datum of appropriate age for the desired results. Whenever available, a time-stratigraphic unit such as a bentonite or a lava flow should be used. Care should also be taken to ensure that the chosen datum horizon does not reflect paleotopography.

3

The Construction of 2-Dimensional Cross Sections

Beginning the construction of a cross section in a new region can be difficult. Most geologists start subsurface sections by comparing wireline logs from local wells to determine the stratigraphy of the area. After general correlations have been determined, the logs can be pinned up along a wall to construct rough cross sections.

    First, a horizontal datum line is established along the wall. Well locations are scaled off horizontally according to the relative distances between wells on the line of section. Each log is positioned over its well location by matching the horizontal datum line and the corresponding elevation or horizon on the log. This procedure is called "hanging" the log on the datum line. Correlatable strata on each log can be marked with colored push pins and the geology can be further highlighted by stretching string or elastic bands between the pins. It is easy to change correlations or to substitute new logs into the diagram. Many rough cross sections can be constructed in this manner while the geologist begins to visualize the geology of the area.

    The first step in drafting a cross section is to determine what the diagram is intended to illustrate. Keeping this purpose in mind, the parameters discussed in chapter two are chosen. Beyond this point, the procedure for plotting data will vary depending on whether the diagram is a structural or a stratigraphic cross section and, to a certain extent, on the type of data to be used.

3.1 Structural Cross Sections

In petroleum geology, most structural cross sections are begun by drawing a horizontal datum line which represents an elevation. Wells may be represented by lines drawn to scale, called sticks, or by well logs reduced to the appropriate vertical scale. In both cases, the horizontal distance between adjacent wells is measured off on the horizontal scale, and the wells are hung on the datum by matching up the datum line with the corresponding elevation in the well (figure 3.1). When sticks are used, it is necessary to scale off the elevations of formation boundaries from the datum. If well logs are used, it is only necessary to mark correlative stratigraphic units in each log. Dipmeter readings and migrated

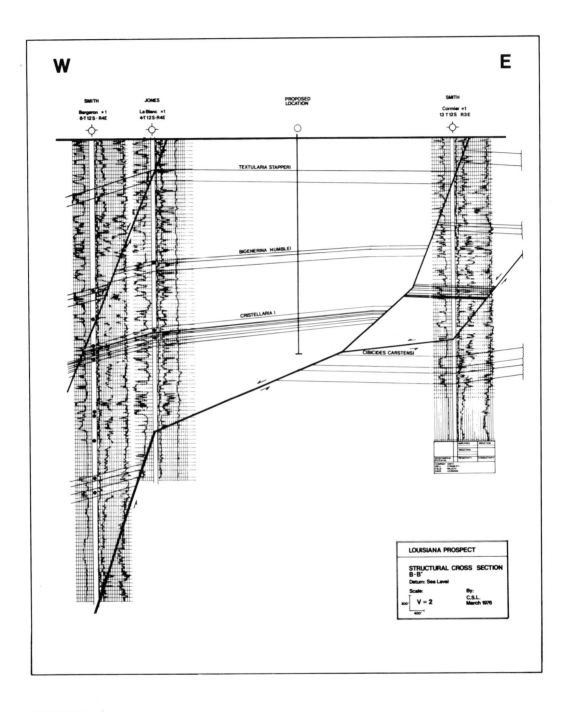

FIGURE 3.1
Structural cross section. Wells are graphically reduced wireline logs.

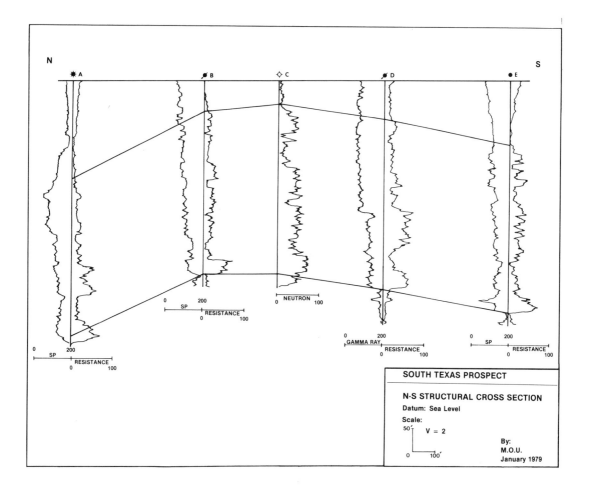

FIGURE 3.2
Wireline log curves which are commonly used in cross sections. Usually only two curves are reproduced: one on the left side and one on the right side of the log print-out. Here, the curves have been redrawn from the original logs.

seismic depths may also be plotted if they are available.

Whenever wireline logs are used, only two of the log curves are usually reproduced: one on the right side and one on the left side of the log print-out. The most commonly used curves are the self-potential or the gamma ray curves on the left and one density or resistivity curve on the right (figure 3.2). Other combinations are possible, depending on what logs are available

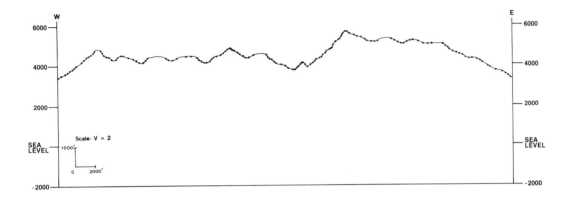

FIGURE 3.3
The construction of a structural cross section showing topography and outcrop geology: plotting the topographic profile from a contour map.

and on what the cross section is intended to show.

Some structural cross sections extend up to the surface of the earth. Such cross sections are often topped by a topographic profile. This profile is derived from a geological outcrop map which shows topography (e.g., most geological quadrangle maps). The generation of the cross section is facilitated by taking the horizontal scale of the diagram to be the same as the scale of the map. The locations of topographic contours and geological contacts can then be transferred directly onto the cross section from the map.

First, draw the line of section on the map. Next, place the base of the drawing paper along the line of section so that the westernmost end of the cross section will be on the left side of the paper. Mark off the points where the paper crosses contour lines, and use the chosen

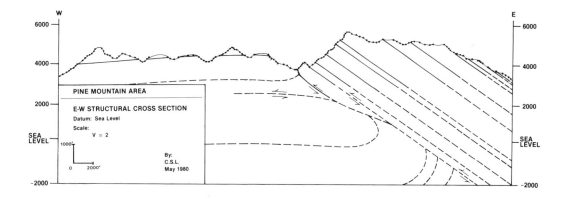

FIGURE 3.4
The construction of a structural cross section showing topography and outcrop geology: plotting geological contacts and structural dips on the topographic profile.

vertical scale to plot the corresponding elevations above each mark (figure 3.3). Connecting the elevation points gives rise to a completed topographic profile.

Next, outcrop data are plotted on the topographic profile. The base of the drawing paper is placed back along the line of section and points are marked where the paper crosses geological contacts, faults, or dip measurements. This information is transferred vertically onto the topographic profile. The accuracy of the plotted positions can be checked by comparison with the geological map.

Dips are drawn in with the help of a protractor (figure 3.4). It is important to remember that dip measurements must be adjusted if the cross section is vertically exaggerated. This is done by solving δ_E in equation 2.3. Graphic solutions are also given in figure 2.11 for values of V up to 15.

3.2 Stratigraphic Cross Sections

FIGURE 3.5
Stratigraphic cross section using both surface and subsurface data. Outcrop information appears in the columnar section at the far left of the diagram.

Stratigraphic cross sections are based on flattened stratigraphic datums. The chosen datum horizon is represented as a straight horizontal line regardless of its actual relief. Since this flattening distorts spatial relationships, topography is not shown.

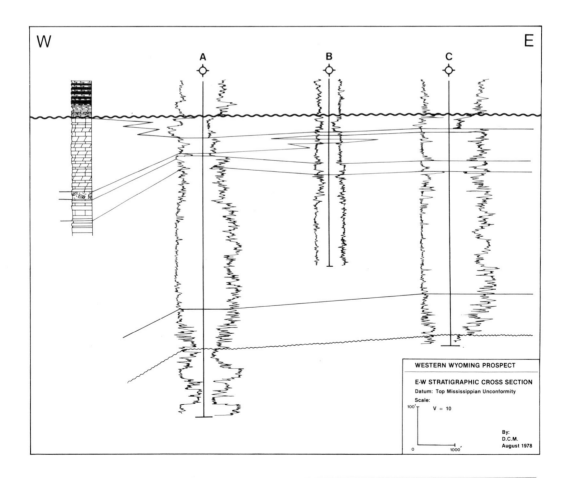

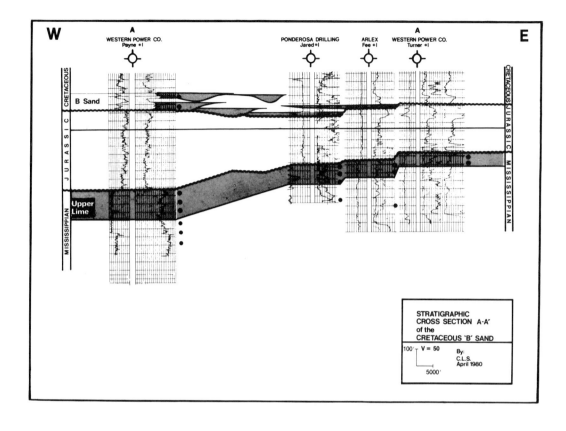

FIGURE 3.6
Stratigraphic cross section based entirely on subsurface well data.

Either surface information, usually in the form of columnar sections, or subsurface information can be used in stratigraphic cross sections (figures 3.5 and 3.6). Both columnar sections and wells are hung on the stratigraphic datum by superimposing the appropriate horizon over the datum line. Where stick wells are used, the stratigraphic intervals to formation boundaries are scaled off. In the case of well logs or of columnar sections it is only necessary to highlight the significant lithological boundaries.

3.3 Correlation and Geological Interpretation

We now have our cross sections laid out. The datum lines are established. Surface and subsurface information have been properly positioned with respect to the datum and to the horizontal and vertical scales. It is best to ink in the "hard" data plotted up to this point in order to avoid erasure while working on the geological interpretation. It may also be helpful to make a copy of the "hard" data for later comparison with the finished cross section.

Correlating data points require the use of geological principles and common sense. One set of data may give rise to more than one acceptable interpretation. In choosing between interpretations, one must first ensure that each approach is reasonable. Scrutinize the data, assumptions, and conclusions. Then make your final choice based on your knowledge of local and regional geology.

Figure 3.7 shows two SP logs, one for Well A and one for Well B. The sandstone shown in each log has been correlated as a discrete formation. In Well A, the sandstone is at a higher elevation and it is water-wet. In Well B, the sandstone is at a lower elevation and is also water-wet, but it contains an oil show. Given this data, there are several possible interpretations: an anticlinal closure between Well A and Well B (figure 3.8); a closure on the upthrown flank of a fault located between the wells (figure 3.9); a closure on the downthrown flank of a fault between the wells (figure 3.10); and finally, a permeability barrier (figure 3.11). The known structural and stratigraphic grain of the area would be used as a guide in determining which of these interpretations is the most likely to be correct.

The accuracy of the interpretation can be checked by comparison with other nearby studies, such as cross sections, geological maps, and geophysical surveys. Specific geological conditions sometimes permit special methods of checking for accuracy. For instance, the concept of volumetric accuracy, based on the principle of conservation of mass, is often used for structural cross sections in areas where concentric folding and thrust faulting have predominated.

The concept of volumetric accuracy holds that, since cross-

FIGURE 3.7
Wireline log curves from two adjacent wells. The logs are hung on a structural datum line.

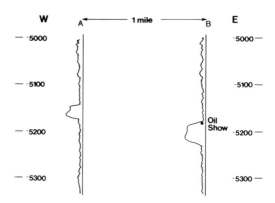

FIGURE 3.8
Interpretation of the data shown in figure 3.7. Anticlinal closure.

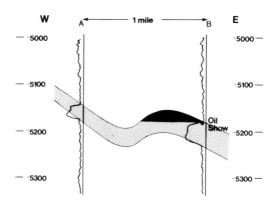

FIGURE 3.9
Interpretation of the data shown in figure 3.7. Closure on the upthrown flank of a fault.

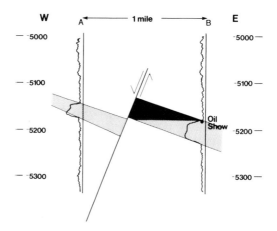

FIGURE 3.10
Interpretation of the data shown in figure 3.7. Closure on the downthrown flank of a fault.

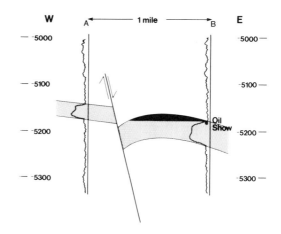

FIGURE 3.11
Interpretation of the data shown in figure 3.7. Permeability barrier.

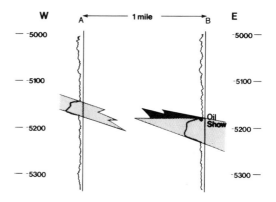

sectional bed length remains constant during concentric deformation, the accuracy of a cross section can be verified by comparing the lengths of several marker horizons. The lengths are measured between two appropriate reference lines. These reference lines are chosen so that they lie in planes where interbed slip has not occurred, such as the axial planes of folds. The lengths of different beds between these reference lines should be equal unless the cross section is cut by a discontinuity. Interpretation of the type of discontinuity will depend on the available data and on local geological trends. For greater detail, the reader should refer to C.D.A. Dahlstrom's article (1969) which describes the use of volumetric accuracy in constructing cross sections through the Alberta foothills region of Canada.

3.4 Completing the Cross Section

The cross section is completed by inking in the final interpretation, and by labeling important features. Conventional colors and symbols are often used to show the geological nature of the features. Some conventions, such as those for sandstone, are more universal than others; the choice itself is less important than consistency once the choice is made. Some common symbols are shown in figure 3.12. Sandstone is often yellow with a stippled pattern; shale is grey with

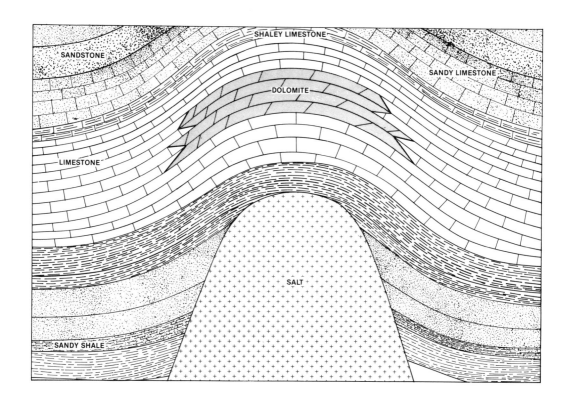

FIGURE 3.12
Common lithological symbols used in cross section construction.

rows of dashed lines; limestone is blue with an orthogonal brick pattern; dolomite is pink, or sometimes blue, with a rhombic brick pattern. Salt is often colored white or orange with rows of small crosses. In special cases where crystalline rocks are shown, they may be colored red. Intermediate lithologies are illustrated by combining the appropriate patterns and by using the color of the predominant lithologic end member.

Pore fluid composition can also be indicated with colors. Gas is shown as red, oil is green, and water is colored blue.

Clear labeling of all important features is essential to an effective cross section. Horizontal and vertical bar scales should also be clearly shown so that the scale is retained if the cross section is photographically reduced or enlarged. In addition, every cross section should include a title, the location and orientation of the line of section, the degree of vertical exaggeration (V); and the author, date, and data sources.

4

3-Dimensional Diagrams

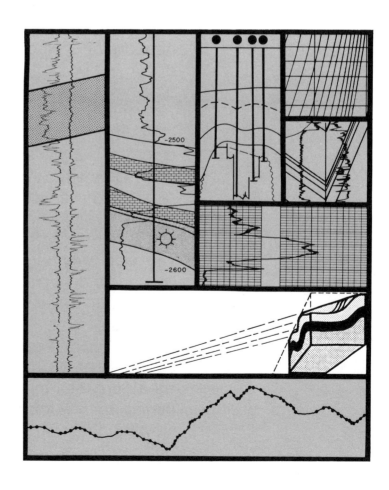

4.1 Fence Diagrams

In petroleum exploration, the development of a prospect is a three-dimensional problem. One of the drawbacks of cross sections is that they are limited to two dimensions. How can we illustrate the three-dimensional geology of an area?

The most common type of diagram showing geological relationships in three dimensions is a fence diagram (figure 4.1). Fence diagrams consist of a three-dimensional network of geological cross sections drawn in two dimensions. The diagram is built on a map base which is seen in plan view.

In constructing a fence diagram, the plane of the map base corresponds to the chosen datum plane, and the well location on the map is taken to be the point where the well intersects the datum plane. The wells are hung on the datum and then individual cross section panels are filled in, beginning at the front of the diagram.

In fence diagrams, one must consider the overall orientation of the figure, bearing in mind that panels oriented parallel to the viewing direction will appear only

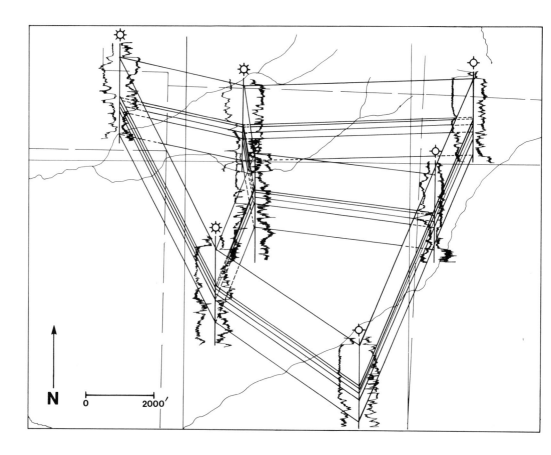

FIGURE 4.1
Fence diagram.

as straight lines. Such panels are usually omitted from the diagram. If important panels are not displayed to the best advantage, the viewing direction can be changed so that the features of interest are presented in the most effective way.

Front panels sometimes obscure parts of rear panels in fence diagrams. Where two panels overlap, the correlations on the front panel are drawn in with solid lines and highlighted with colors, while the correlations on the overlapped part of the rear panel are simply dashed in.

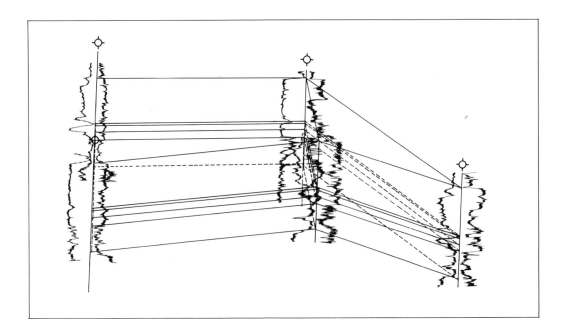

FIGURE 4.2
Fence diagram. The vertical scale is large relative to panel spacing, resulting in large parts of the rear panels being obscured.

When the vertical scale is too large or the spacing of sections is too close, very large parts of the rear panels may be obscured (figure 4.2). Usually, this problem can be minimized if the vertical scale and the panel spacing are carefully chosen. However, it is not always possible to change the scale or spacing without losing important details illustrated by the diagram. In such a case, one can construct an isometric projection of the fence diagram (figure 4.3b).

The map base in an isometric projection is shown as if it was turned at an angle and tilted toward the viewer. The conversion is actually a transformation from orthogonal to nonorthogonal axes. A rectangular map base will become a parallel-

FIGURE 4.3
Transformation of a fence diagram (a) into an isometric projection (b). Only lines parallel to the North-South and East-West axes are undistorted in the isometric projection.

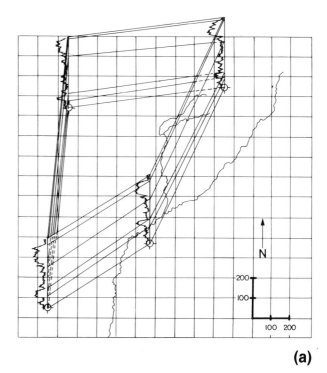

(a)

(b)

3-DIMENSIONAL DIAGRAMS ■ **51**

ogram in an isometric projection (figures 4.3a and 4.3b). All lines that were parallel to the original horizontal and vertical axes remain parallel to the corresponding new axes, and the scale along these lines also remains unchanged. However, lines which were not parallel to the original axes will be distorted. Points along such lines must be transferred onto the projection by coordinates relative to the new axes. A grid system is helpful in this process.

4.2 Block Diagrams

Another type of diagram illustrating three-dimensional geological relationships is a block diagram (figure. 4.4). This type of figure is a two-dimensional representation of a rectangular block. Two intersecting cross sections often form the sides of the block; the top of the block shows either a mapped surface or relief on the uppermost geological boundary shown in the cross sections. The block may be rotated to achieve the most effective viewing angle, and it may be drawn with or without the use of perspective.

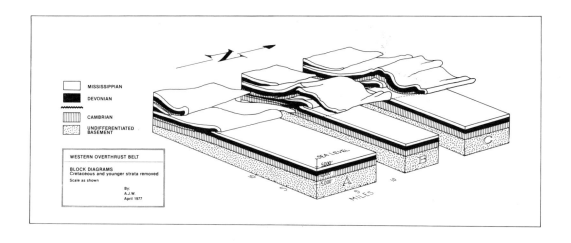

FIGURE 4.4
Block diagrams. (Courtesy of Al Witherspoon, Mobil E&P, Denver).

Isometric block diagrams do not employ perspective. Opposite sides of the block remain parallel, making the distant end of the diagram appear to be larger (figure 4.5). The top of this type of block diagram may be given any degree of tilt toward the viewer. A large amount of tilt emphasizes features on the top surface of the block; a smaller degree of tilt brings out details on the sides of the block. Scales along the x, y, and z axes need not be equal.

The scale along the sides of the figure (y) is sometimes chosen to be smaller than the scale across the front and back edges (x), thus creating an illusion of perspective.

Perspective can be used in constructing block diagrams (figure 4.6). In perspective blocks, all parallel lines with a component in the y direction converge to a vanishing

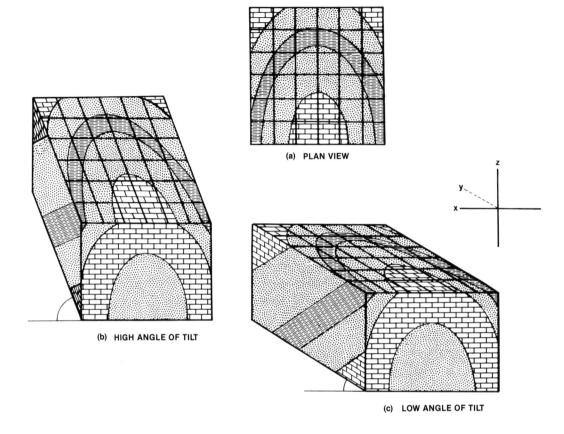

FIGURE 4.5
Isometric block diagrams.

point on the horizon. Sets of parallel lines with different orientations converge to different vanishing points. The cross section on the front face of the block is usually left undistorted by perspective. It is possible both to rotate the block about the z axis or to display it at different elevations relative to the horizon. Again, the choice of block orientation is made to emphasize the important features of the diagram.

Since their complexity inhibits detailed illustration, block diagrams

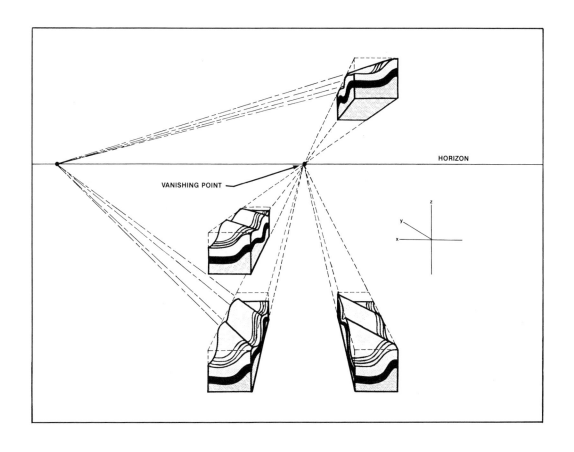

FIGURE 4.6
Perspective block diagrams.

are usually of a diagrammatic nature. They make effective displays for articles, presentations, and exhibits. However, they are difficult to construct. Fence diagrams are relatively easy to construct and can be made to show greater detail than block diagrams. Thus, in petroleum exploration, fence diagrams are used more often than block diagrams when illustrating the three-dimensional geology of an area.

5
Computer-Drawn Cross Sections

In some cases computers can facilitate the long process of cross section construction. A computer can generate cross sections rapidly and inexpensively. The information required is the same as for drafting a cross section by hand, however, the data must be translated into a code which the computer can read. Coded data and programs are convenient to store and to retrieve, permitting reproduction of old cross sections at any time. Perhaps most importantly, parameters are easily changed until the most effective diagram is obtained. Where needed, the computer programs themselves can always be revised or updated. Computer-generated cross sections are particularly useful as working diagrams used during the problem-solving stage.

    It is important to be aware of the limitations of the computer. Parameters and limitations will vary depending on the type of computer and on the programs available for use. Properly instructed, the computer can incorporate the personal knowledge and experience of the geologist into the cross section it generates. However, computers do not make any interpretation beyond that made by the geologist when the data input was selected.

For instance, some cross section programs will not draw faults. In such cases, segments containing faults should be left blank by the computer and correlated by hand if the fault is desired on the cross section. Previously unknown fault locations may show up as distortions in the computer-drawn section (figure 5.1). The geologist should examine the data surrounding such an anomaly and, if necessary, revise the cross section by hand.

There are several types of computer sections. In the simplest case, the computer can plot stick wells without showing correlation lines. The only input needed by the computer for this type of section is the specification of the scale, distances between wells, elevation of the wellhead, total depth, and the elevation of the datum in each well. If the wells are to be represented by logs (figure 5.2), then the computer requires the logs in an acceptable digitized form. When connection of geological boundaries is desired (figure 5.3), the boundary elevations in each well must also be specified.

Using the same method, many computers can draw fence diagrams from any viewing angle (figures 5.4a, 5.4b, and 5.4c). The desired angle must be given, in addition to all well information and datum and geological boundary elevations.

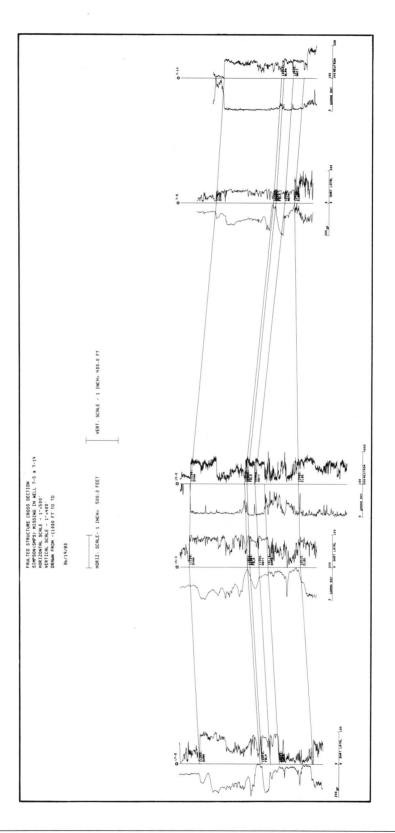

FIGURE 5.1
Computer cross section showing distortion due to a fault. (Courtesy of Tom Deubel, Mobil).

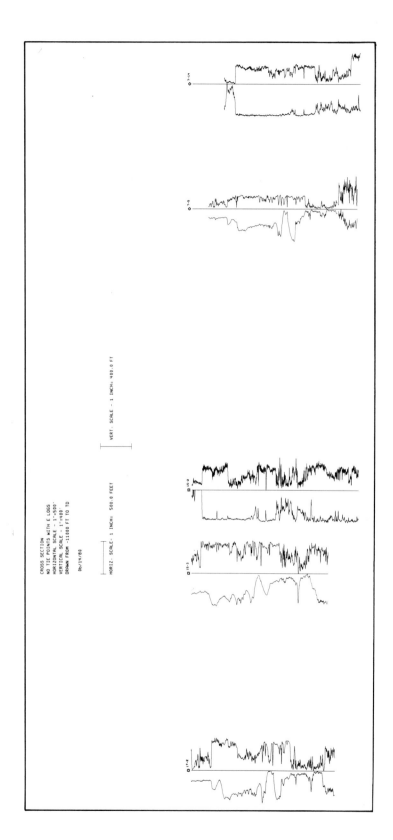

FIGURE 5.2
Computer plot of well locations using wireline logs. (Courtesy of Tom Deubel, Mobil).

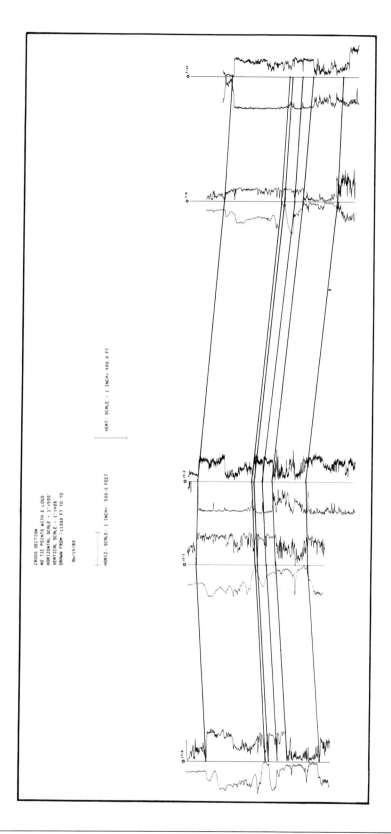

FIGURE 5.3
Well locations from figure 5.2 with geological boundaries connected by the computer. (Courtesy of Tom Deubel, Mobil).

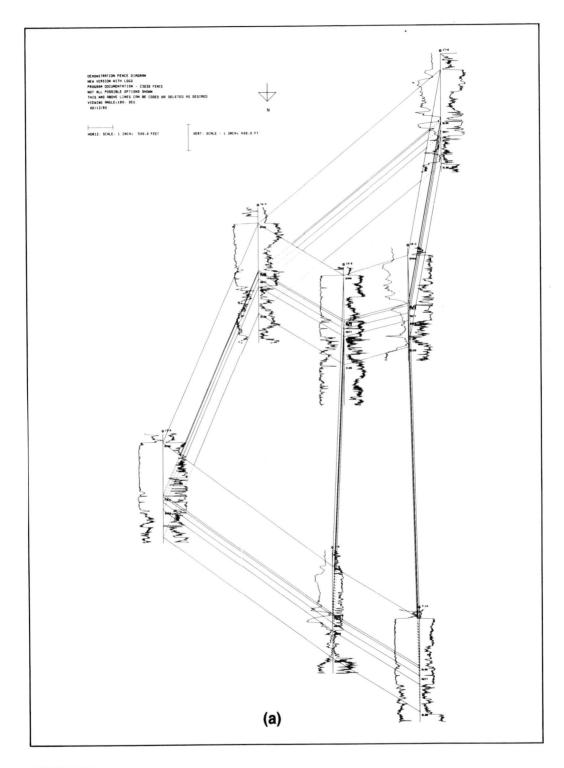

FIGURE 5.4a
Computer-drawn fence diagram seen from an 180° viewing angle.
(Courtesy of Tom Deubel, Mobil).

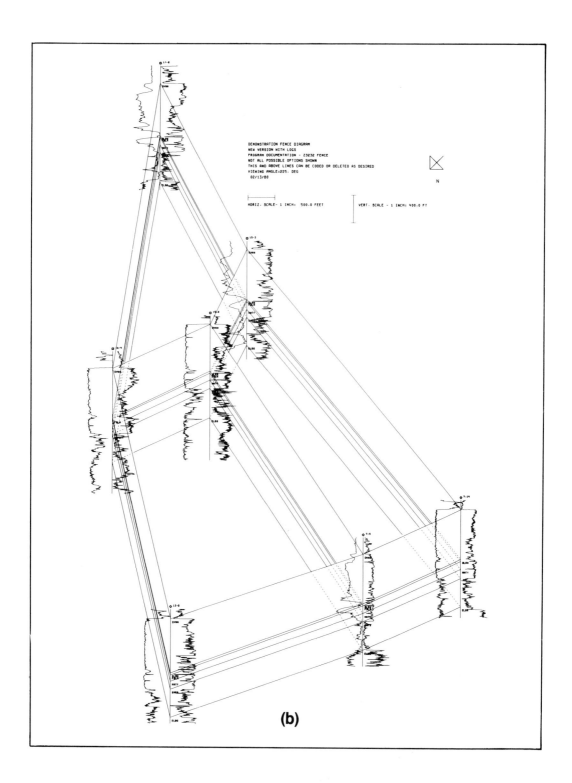

FIGURE 5.4b
The computer-drawn fence diagram illustrated in figure 5.4a is seen here from a 225° viewing angle. (Courtesy of Tom Deubel, Mobil).

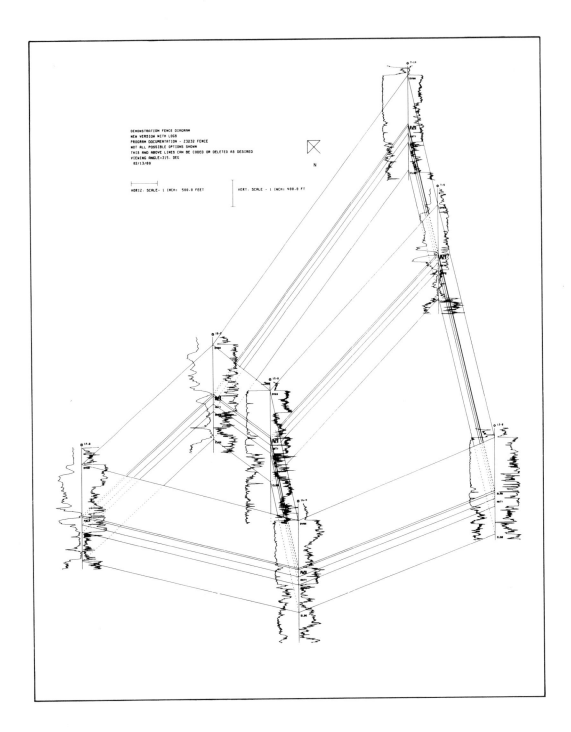

FIGURE 5.4c
The computer-drawn fence diagram illustrated in figures 5.4a and 5.4b is seen here from a 315° viewing angle. (Courtesy of Tom Deubel, Mobil).

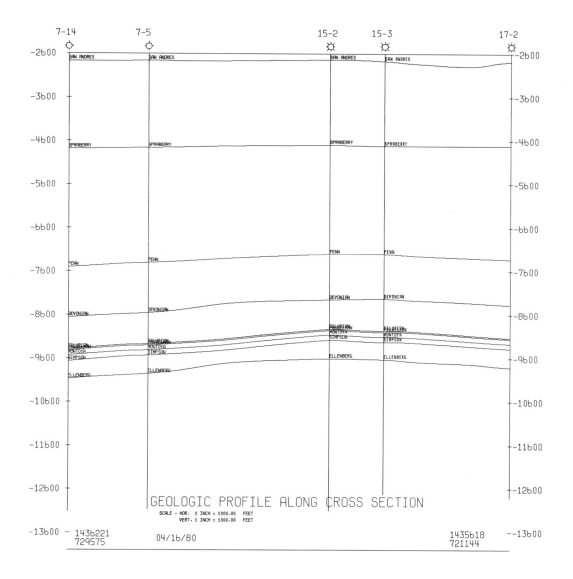

FIGURE 5.5
Gridded profiles stacked to form a cross section. (Courtesy of Tom Deubel, Mobil).

Since fence diagrams are somewhat time-consuming to construct, the advantages of computer application are evident. Using the computer, it is easy to experiment with different parameters until the most effective diagram is obtained.

Gridded profiles are a second category of computer-generated cross section (figure 5.5). Each subsurface geological boundary is treated independently in the gridded profile approach. First, the computer locates all data points which give the elevation of the boundary over the region in question. Next, the computer fits a grid over the available regional data. The geometric characteristics of the grid vary from program to program, depending on the desired time and storage efficiency. Common grid types are square (figure 5.6) or triangular. The size of the grid cell is an important variable. Too large a grid spacing will cause smoothing which may eliminate significant details. When the grid is too small, large clusters of empty grid squares will occur between data points, which may result in inaccurate calculations for grid point values near the centers of the empty clusters. The optimum grid spacing depends on the average distance between grid points, and on the type of grid being used. The computer can accomodate a very large, but finite, number of grid cells.

After the chosen grid has been fit over the data points, the computer calculates a weighted average value for each point of intersection on the grid. This is done by searching the adjacent grid cells

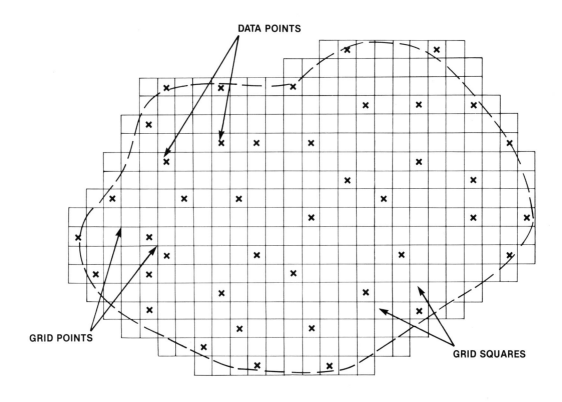

FIGURE 5.6
Computer gridding of random data points. (Courtesy of Mobil Mapping and Navigation).

for neighboring data points. The values at these neighboring data points are weighted according to their distance from the grid point, and all of the weighted values are averaged to give an estimated value for the grid point. Where data is sparse, each grid point may be smoothed by an average of the neighboring grid point values.

The next step can be to contour the calculated grid point values (figure 5.7). This is done one grid cell at a time by fitting a polynomial of given order to the grid square.

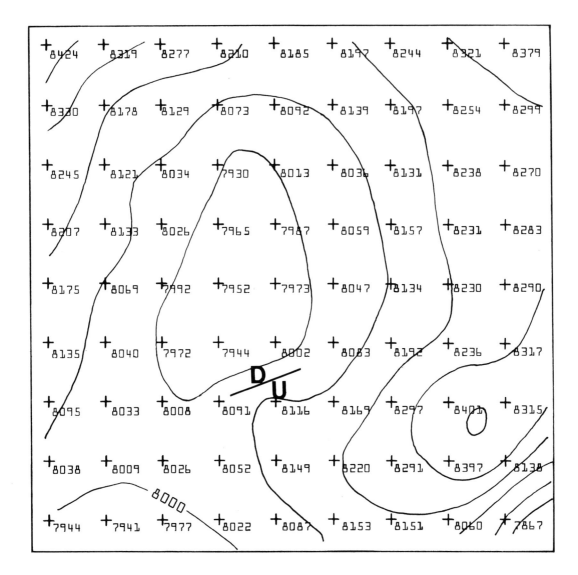

FIGURE 5.7
Computer contouring of gridded data. (Courtesy of Mobil).

The computer will draw profiles along any chosen line of section on the contoured gridded surface. Cross sections are formed by stacking two or more different profiles made along the same line of section.

Handling of discontinuities such as faults will vary from program to program. Often, if the surface to be

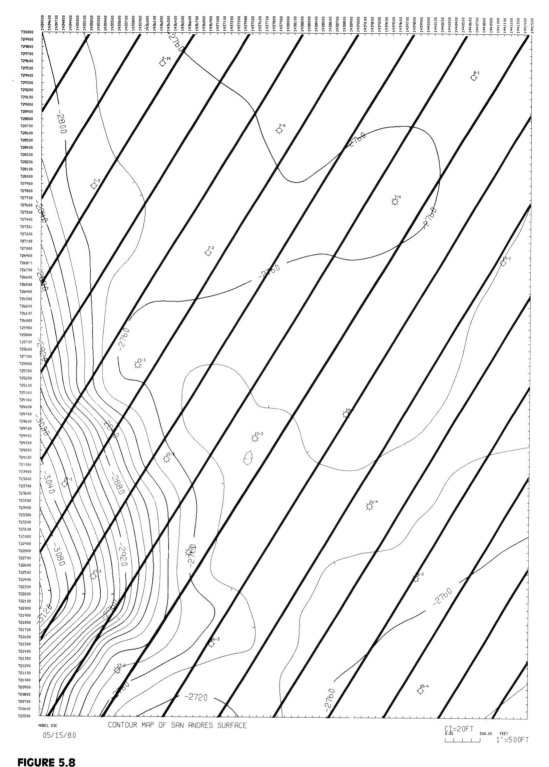

FIGURE 5.8
Lines of section on a contoured gridded surface. Using gridded data, it is possible to choose lines of section without regard to the location of individual wells.

gridded is offset by a fault, the location and orientation of the fault trace must be included in the data input. The computer then will not average data points, nor will it contour, across the fault. If the location of an existing fault is not indicated, the computer may smooth across the fault.

Gridded profiles are fairly accurate in areas of dense data control. However, the accuracy breaks down in areas of poor control. Where sparse data is relatively evenly spaced, one solution is to choose a large grid size which will smooth the data to reveal only regional structural trends. When data points are clustered, it may be necessary to smooth or even to omit the data in the intervening areas of poor coverage.

Several advantages are evident in the gridded profile method of cross section generation. First, a straight line of section can be chosen with or without regard to the location of wells. Second, any number of arbitrarily chosen profiles or cross sections can be made quickly over the same region. For instance, one could generate a parallel series of straight-line sections over a region (figure 5.8). However, one must be careful to check the grid size and the extent of data control since these factors will affect the accuracy of the gridded profile.

6

Review

We have defined a cross section as a profile showing geological features in any vertical plane through the earth. This type of diagram can be divided into two primary categories: structural and stratigraphic. Structural cross sections are referred to elevation datums and show present-day geological structure. Stratigraphic cross sections are referred to flattened stratigraphic datums and show structural relationships which existed at the time the datum boundary was formed. Important elements which make up a cross section are the line of section, data points, datum line, and horizontal and vertical scales. These elements are chosen to best illustrate the geological features of interest. The general procedure for plotting cross sections is outlined in table 6.1.

It is most important to be aware of distortions caused by nonlinear lines of section, vertical exaggeration, and borehole deviation. These distortions may have significant impact on the quantitative features in a cross section. It is also important to be aware of potential sources of error. All data should be carefully checked for accuracy.

Cross sections may be used in the construction of diagrams which illustrate three-dimensional geological relationships. Two of the most

TABLE 6.1 Procedure for Plotting Cross Sections.

1. Choose what the cross section is to illustrate
2. Choose the type of cross section and datum
   A. Structural (elevation datum)
   B. Stratigraphic (flattened stratigraphic datum)
3. Choose a line of section
4. Choose data points
5. Choose horizontal and vertical scales
6. Plot available data
7. Correlate data

common three-dimensional figures are fence diagrams and block diagrams.

Finally, computers can be used to generate cross sections and fence diagrams. There are two types of computer cross sections: those plotted directly from input data and those generated from gridded data. The most important advantage of computer-drawn diagrams is that parameters can be easily changed until the most effective diagram is obtained. It is important to be aware of the limitations of the machine, and to work within those limitations. When properly utilized, computers are a valuable tool for the construction of cross sections.

7 Supplemental References

Bishop, Margaret S., 1960, Subsurface mapping: New York, John Wiley and Sons, Inc., p. 15-21.

Dahlstrom, C.D.A., 1969, Balanced cross sections: Canadian Journal of Earth Sciences, v. 6, p. 743-757.

Leroy, L.W., and Low, Julian W., 1954, Graphic problems in petroleum geology: New York, Harper and Row, p. 15-30.

8
Questions and Exercises

Questions

1.
What type of cross section would you construct if:

(a) You were interested in the relationship of subsurface features 1 million years ago, and;

(b) you wanted to see how these features changed over a large region, and;

(c) the features are thin and have a very gentle dip?

2.
What type of cross section would you construct if:

(a) You wished to know if a present-day surface feature was reflected in a similar subsurface structure, and;

(b) the area you were interested in was fairly small, and;

(c) it was important that the section not distort the thickness of the strata it depicted?

3.
What are common sources of error in cross sections and how are they avoided?

4.
(a) Using figure 2.11 or equation 2.3 fill in the blanks in this table.

	VERTICAL EXAGGERATION	TRUE DIP	EXAGGERATED DIP
1a.	7	65°	___
b.	7	80°	___
2a.	4	___	71°
b.	4	___	78°
3a.	___	4°	35°
b.	___	6°	47°
4a.	5	___	32°
b.	5	14°	___

(b) What general rule can you observe in 1a and b?

(c) ... in 2a and b?

(d) What can you observe in 3a, b and 4a, b?

5.
What are the benefits of a computerized cross section, and what are its limitations?

6.
Once you have decided what type of cross section you need, what is the construction procedure?

Exercise 1

Information is given below for four wells and one columnar section in northwestern Montana. Use the information to construct a stratigraphic cross section on top of the Mississippian. You will need to know that the evaporitic Potlatch Formation is the eastern facies equivalent of the Wabamun Dolomite. All distance measurements are given in feet.

Well A

ELEVATION:	4291 (DF)
TOTAL DEPTH:	6642
LOG PICKS:	
Mississippian (unconformity)	4622
Sun River	4622-4780
Lodgepole/Mission Canyon	4780-5850
Devonian (unconformity)	5850
Potlatch	5850-6070
Nisku	6070-6090
Duperow	6090-
REMARKS: D&A	

Well B

ELEVATION:	4347 (DF)
TOTAL DEPTH:	4692
LOG PICKS:	
Mississippian (unconformity)	2872
Sun River	2872-3023
Lodgepole/Mission Canyon	3023-4050
Devonian (unconformity)	4050
Three Forks	4050-4075
Potlatch	4075-4295
Nisku	4295-4338
Duperow	4338-
REMARKS: D&A	

DISTANCES BETWEEN DATA POINTS

Outcrop to Well A:	26500 ft E
Well A to Well B:	21000 ft E
Well B to Well C:	10500 ft E
Well C to Well D:	10500 ft E

Well C

ELEVATION:	3750 (DF)
TOTAL DEPTH:	3450
LOG PICKS:	
Mississippian (unconformity)	1910
Sun River	1910-1980
Lodgepole/Mission Canyon	1980-3050
Devonian (unconformity)	3050
Three Forks	3050-3115
Potlatch	3115-3290
Nisku	3290-3350
Duperow	3350-
REMARKS: 200 BOPD	

Well D

ELEVATION:	3714 (DF)
TOTAL DEPTH:	3590
LOG PICKS:	
Mississippian (unconformity)	1950
Lodgepole/Mission Canyon	1950-3070
Devonian (unconformity)	3070
Three Forks	3070-3158
Potlatch	3158-3298
Nisku	3298-3350
Duperow	3350-
REMARKS: D&A	

Top of Mississippian

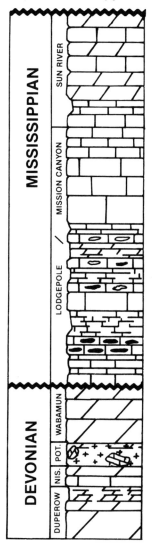

Worksheet—Exercise 1

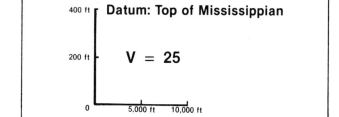

Datum: Top of Mississippian

V = 25

Exercise 2

The following four well logs have been hung on a structural datum. Interpret the geological relationships shown in each by drawing a structural cross section through the logs.

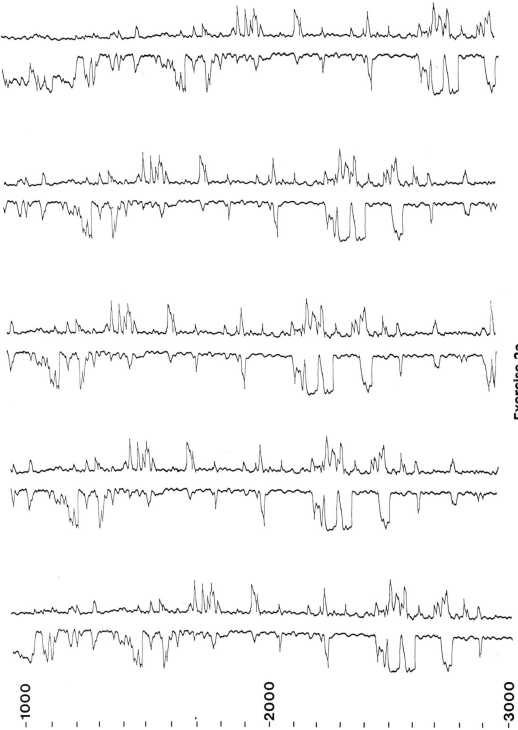

Exercise 2a

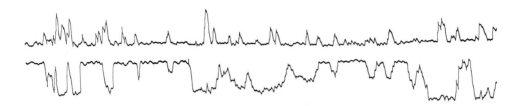
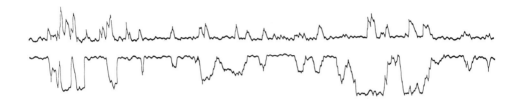
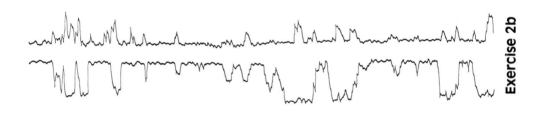

Exercise 2b

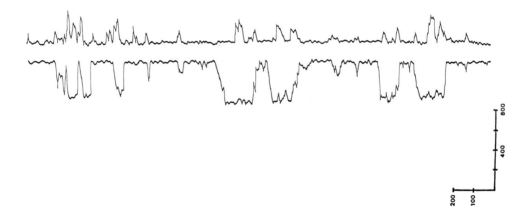

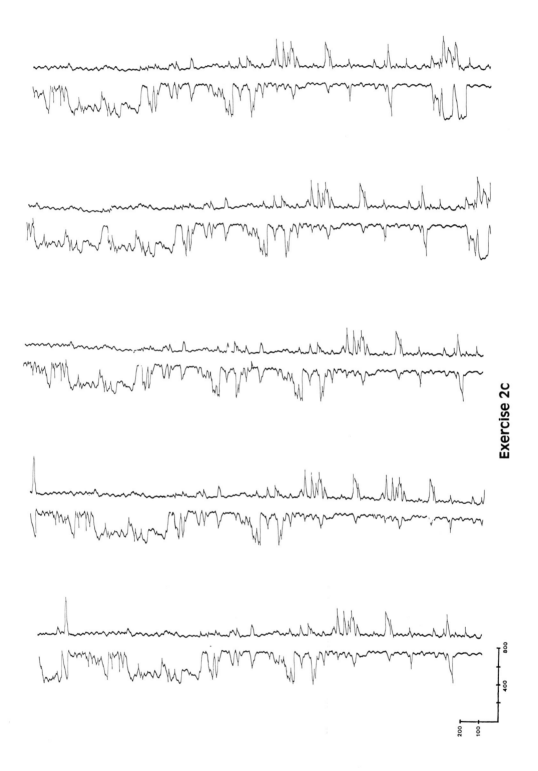

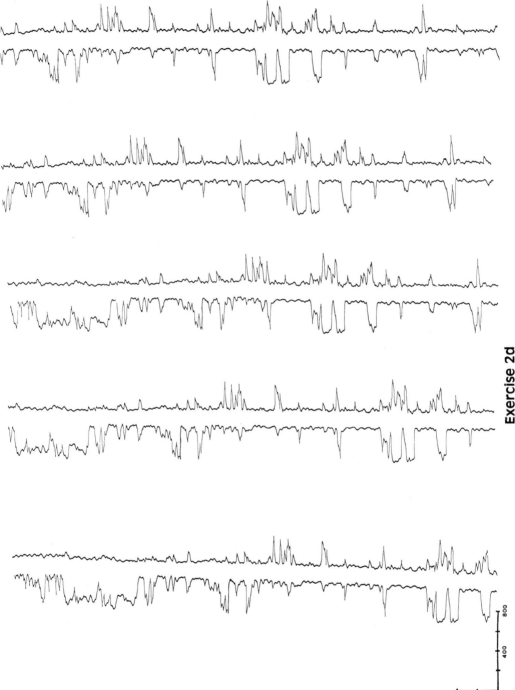

Exercise 2d

Exercise 3

Information is given below for four wells. Use the information to construct a structural cross section. All distance measurements are given in feet.

Datum: −5,000 ft
Scale: 100 ft / 500 ft, V = 5

DISTANCES BETWEEN WELLS

Unit 1A to Unit 2:	3850 ft SE
Unit 2 to Unit 3:	2250 ft SE
Unit 3 to Unit 4:	2400 ft SE

Well—Unit 1A

ELEVATION:	234 (KB)
TOTAL DEPTH:	6154
DRILLERS' PICKS:	
A Sand	5634-5724
B Sand	6102-6146
REMARKS:	D&A

Well—Unit 2

ELEVATION:	127 (KB)	
TOTAL DEPTH:	6127	
LOG PICKS:		DIPMETER READINGS:
A Sand	5317-5349	10° SE
B Sand	5725-5770	15° SE
C Sand	Absent	—
REMARKS:	Possible fault at 6067 ft	
	Gas @ 5317-5377; 5725-5757 ft	

Well—Unit 3

ELEVATION:	252 (KB)	
TOTAL DEPTH:	6212	
LOG PICKS:		DIPMETER READINGS:
A Sand	5617-5712	30° SE
B Sand	Absent	—
C Sand	6075-6162	7° SE
REMARKS:	Gas in C Sand; G/W contact at 6102 ft	

Well—Unit 4

ELEVATION:	398 (DF)	
TOTAL DEPTH:	6400	
LOG PICKS:		DIPMETER READINGS:
A Sand	5518-5598	7° SE
B Sand	5923-5961	7° SE
C Sand	6281-6370	7° SE
REMARKS:	Slight show of Gas at 5518 ft	

Exercise 4

The Wyckoff Gas Field, located in Steuben County, N.Y., produces from Onondaga Limestone and/or Oriskany Sandstone. The Onondaga forms a thick biohermal reef over part of the field. Only the porous core facies is productive in the reef section (see map). A deep-seated down-to-the-southwest fault extends upward along the southwest flank of the reef. Oriskany production is from a small anticline on the upthrown side of the fault.

Elevations and marked logs are provided for 6 wells in the Wyckoff Field. Use this information to construct a northeast-southwest structural cross section from the Richards well to the Dibble well, showing the interval from top of Onondaga to bottom of Oriskany.

Wyckoff Reef Gas Field

WELL	ELEVATION
CORNELL	2257'
DIBBLE	2098'
GUILD	2037'
CHASE	2206'
BANKS	2182'
RICHARDS	2066'

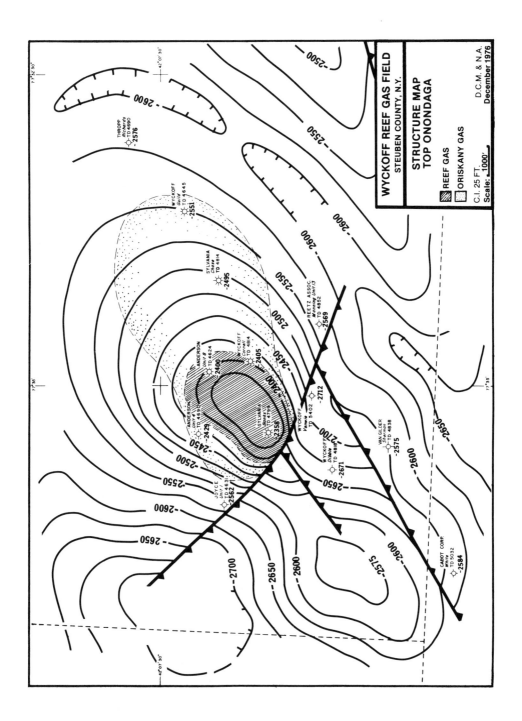

Exercise 4

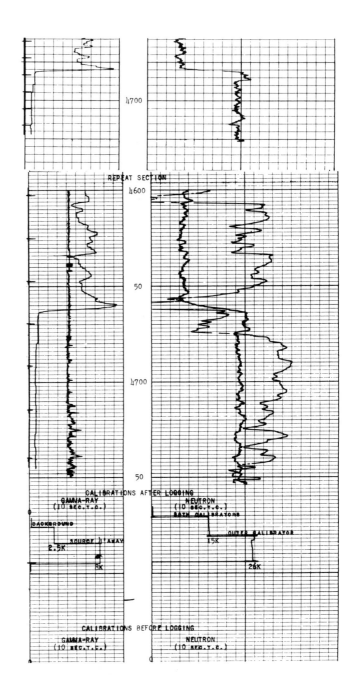

**WYCKOFF DEV. CO.
Cornell
El. 2257'**

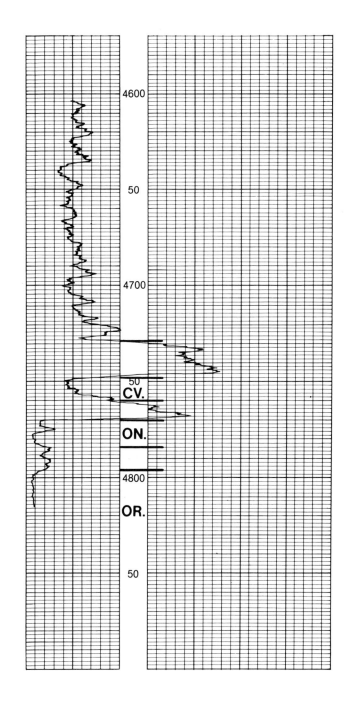

**WYCKOFF DEV. CO.
Dibble
El. 2098′**

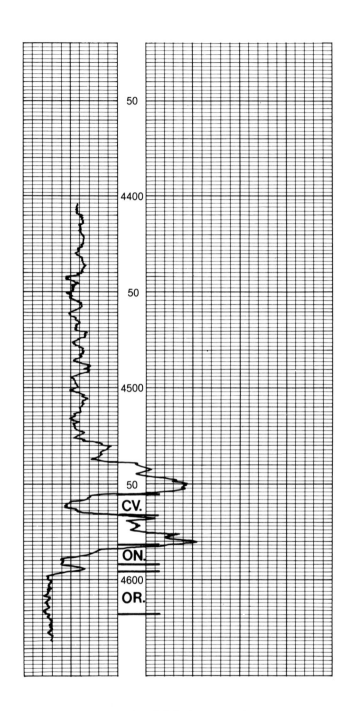

**WYCKOFF DEV. CO.
Guild
El. 2037'**

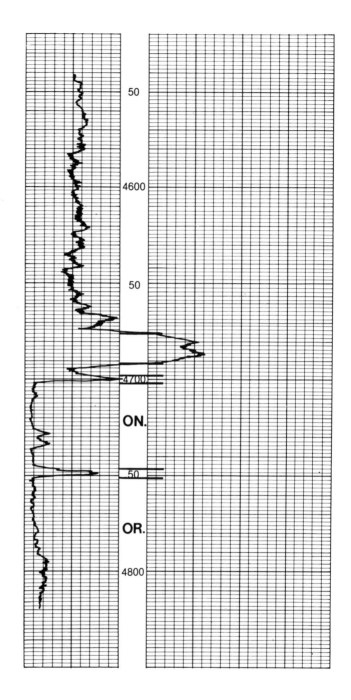

SYLVANIA
Chase
El. 2206'

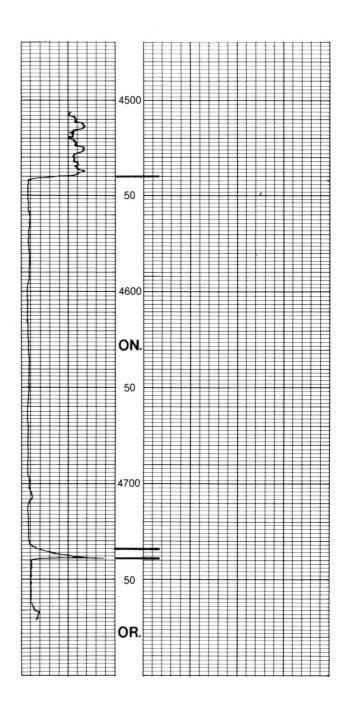

SYLVANIA
Banks
El. 2182'

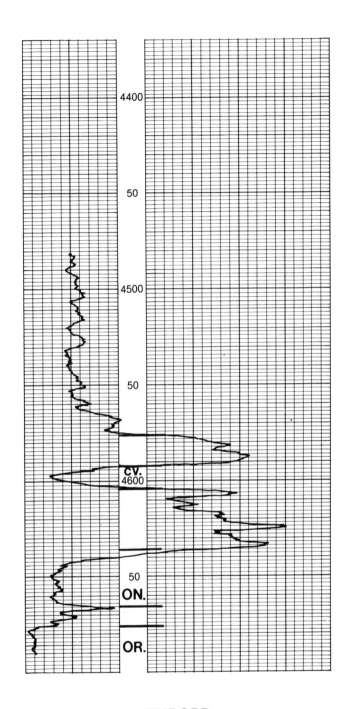

THROPP
Richards
El. 2066'

9
Solutions

Solution 1

The fact that you are interested in paleo-structural relationships rather than current ones dictates the choice of a stratigraphic section. These sections are not tied to present-day topography or to a fixed elevation, but to a particular stratigraphic horizon. The horizon may be tilted, folded, faulted or eroded at the present time, but the stratigraphic section visualizes it as if it was flat. This makes it much easier to understand past geological conditions and can be helpful in locating areas where hydrocarbons might have formed or been trapped.

In this case, the fact that the features are not dramatic together with the vast horizontal scale of a regional section strongly indicates the use of vertical exaggeration. Without multiplying vertical distances the features of this section would be so thin that the section would show virtually nothing. By enlarging the vertical scale several times, beds can be enlarged, dip angles steepened, and trends accentuated. This creates distortion, but given the nature of the section this is justified by the benefits of greater structural clarity.

Solution 2

The need to depict current structures—including topography—dictates the use of a structural cross section. This type of section is tied to actual elevation and represents as accurately as possible the folds, faults, and unconformities of an area. It is intended to reveal subsurface structures that might, for instance, form hydrocarbon traps. In our specific example it is used to verify the presence or absence of a surface feature at depth. If the feature is a dome or an anticline—and if it does exist at depth—then it might commend itself as a drilling prospect.

In this case, the structures of interest are not dwarfed by the area covered in the section, so vertical exaggeration is not necessary. By matching the vertical scale to the horizontal scale actual spatial relationships can be preserved. Because it is important that the features of your section not be distorted, it is preferable not to use any vertical exaggeration.

Solution 3

Constructing cross sections is an inferential process, and the results are largely speculative. Our perceptions of the subsurface are usually based on limited data which often lend themselves to more than one interpretation. Perfect accuracy is unlikely especially with complex structures, and minor errors are therefore inherent in the process. It is especially important, however, to avoid large errors which could detract unnecessarily from the accuracy of a section. Errors of this type usually fall into three categories:

(a) Faulty data—the gathering of all data is subject to error. Sources, especially older ones, should be screened on the basis of reliability. Common errors in older data include incorrect well locations and elevations. Hard data—like logs—are preferable to more subjective information like drillers' picks.

(b) Faulty interpretation of data—the conclusions of previous geologists may be inaccurate. Misinterpretations can be perpetuated if prior work on an area is accepted without question.

(c) Insufficient data—wells may be spaced too widely to generate a reliable section. Structural details may be glossed over if supplementary data (seismic, etc.) is not available.

Solution 4

(a) Missing figures in table:

    1a. Exaggerated Dip = 86°
      b. Exaggerated Dip = 89°
    2a. True Dip = 36°
      b. True Dip = 50°
    3a. Vertical Exaggeration = 10
      b. Vertical Exaggeration = 10
    4a. True Dip = 7°
      b. Exaggerated Dip = 51°

(b) Vertical exaggeration tends to decrease the angular difference between two steeply dipping surfaces. In 1a and b the difference in dip has decreased from 15° (true) to only 3° (depicted).

(c) The same effect is evident in this example, too. However, because the vertical exaggeration is much less and the true dips are not quite so steep the effect is less pronounced. In 1a and b the difference between the true dips is five times as great as the difference between the exaggerated dips. However, in 2a and b the difference between the true dips is only two times the difference between the exaggerated dips.

(d) The effect of vertical exaggeration is just the opposite on gently angled strata. Minor differences at low dip angles are greatly increased in cross sections that are vertically exaggerated. A 2° difference in true dip in 3a, b is magnified to a 12° difference in exaggerated dip; and a 7° difference in 4a, b magnifies into a 19° difference. Vertical exaggeration is therefore more useful in distinguishing between slightly dipping structures than between sharply dipping structures.

Solution 5

In situations where large quantities of data must be sorted and arranged, a computer can frequently do the job faster than a person can. The preliminary aspects of constructing a section—organizing data, hanging logs and sticks on the datum line, and making initial correlations—are time consuming, and can be done quickly by a properly programmed computer. Using a computer it is also possible to vary the parameters of a cross section until the most effective diagram is obtained.

However, the computer cannot make value judgements. Limitations arise because of the intuitive element in interpretation. For instance, the factors that suggest a fault or an unconformity to a geologist are often difficult to quantify—and they are even more difficult to incorporate into a computer program. There is still no substitute for human expertise.

Solution 6

Based on what you want to know, you will choose a line of section that best reveals the features you are interested in. If you are constructing a structural cross section which includes surface topography, you will now add information from a contour map. By matching the scale of the section to that of the map it is possible to transfer elevations directly from one to the other. Elevation points are then connected to form a topographic profile. Using the same procedure geological information can be transferred.

Next, a horizontal datum line is established. It will extend across the diagram and all depth data will be keyed to it. With structural cross sections the datum line is an elevation; with stratigraphic cross sections it is a prominent stratum or other noteworthy plane. Such a plane must be uninterrupted over the full area depicted in the section, and it must not reflect any features of paleotopography. Moreover, the choice must be appropriate for the purpose of each individual cross section.

Data, usually in the form of logs or stick wells, are now sequenced and "hung" on the datum line. The logs or stick wells are scaled off at distances corresponding to actual well spacing. Once each column of data has been keyed to the horizontal datum, initial correlation can begin. As identical horizons in each chart are connected, discontinuities may emerge. The resolution of these questions will take place during the final stages of interpretation.

Once the hard data has been organized and positioned it is a good practice to ink it in before starting the final stage of interpretation. With that done, it is possible to rework the section many times without disturbing the basic data. Specific instructions for the interpretation phase are difficult to give. Each situation is different. Experience and intuition play a big role in filling in the geology between wells.

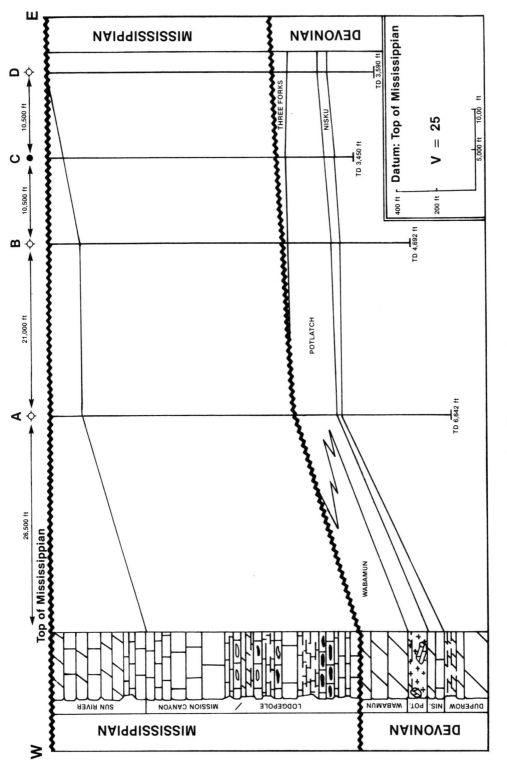

Solution—Exercise 1

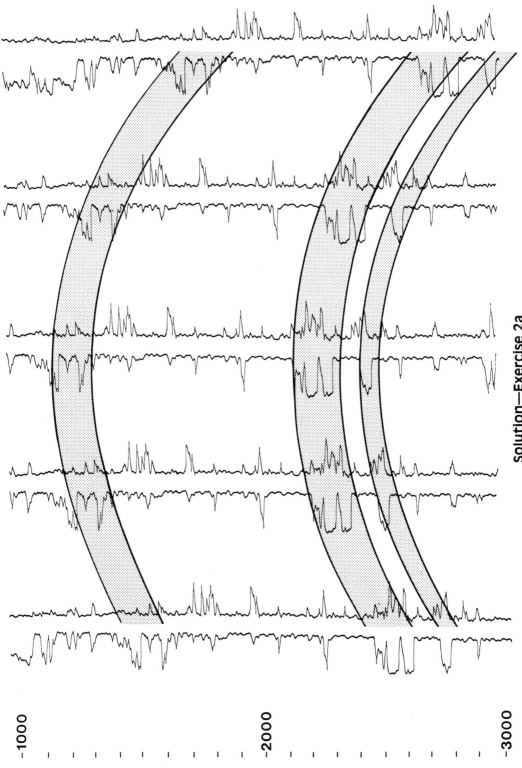

Solution—Exercise 2a

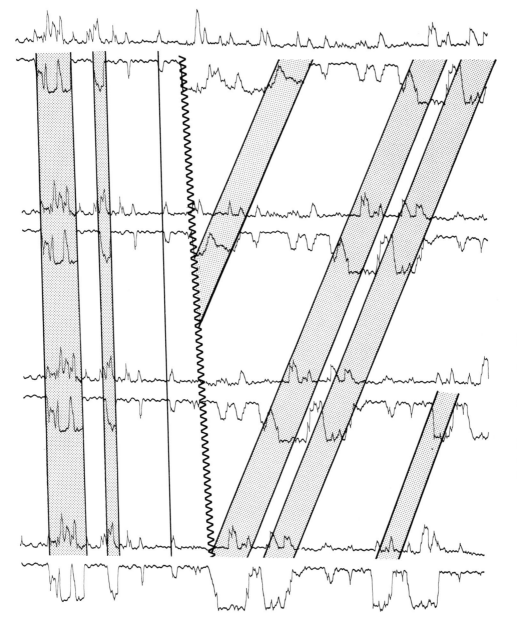

Solution—Exercise 2b

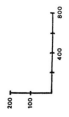

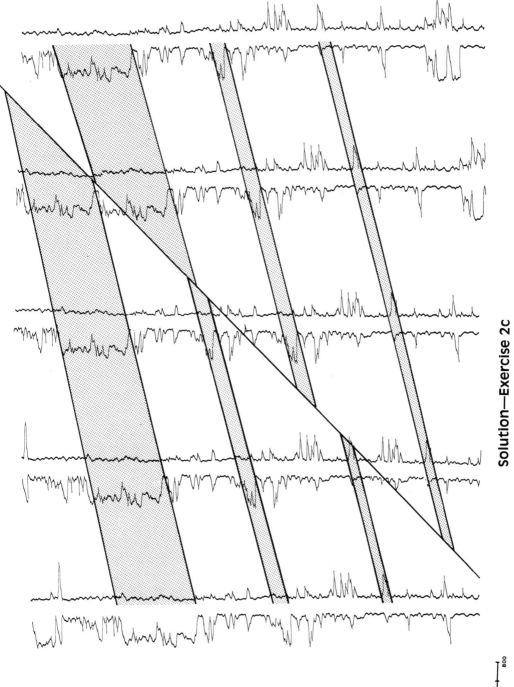

Solution—Exercise 2c

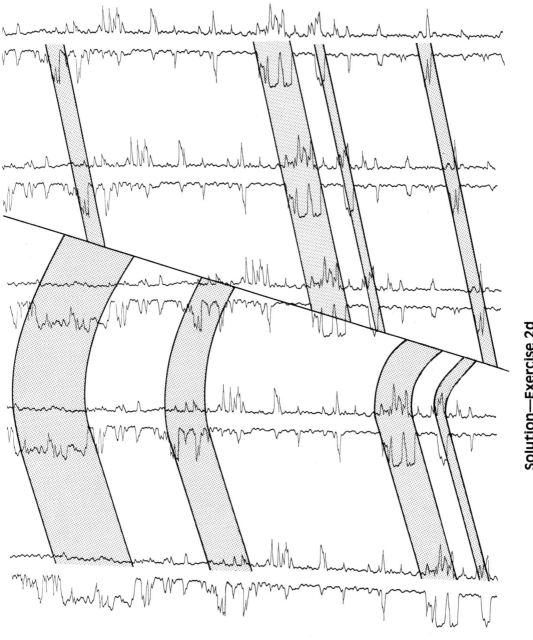

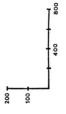

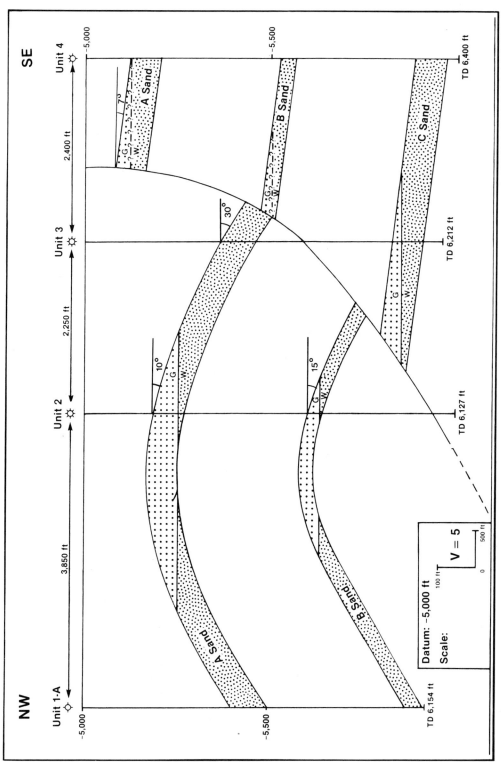

Solution—Exercise 3

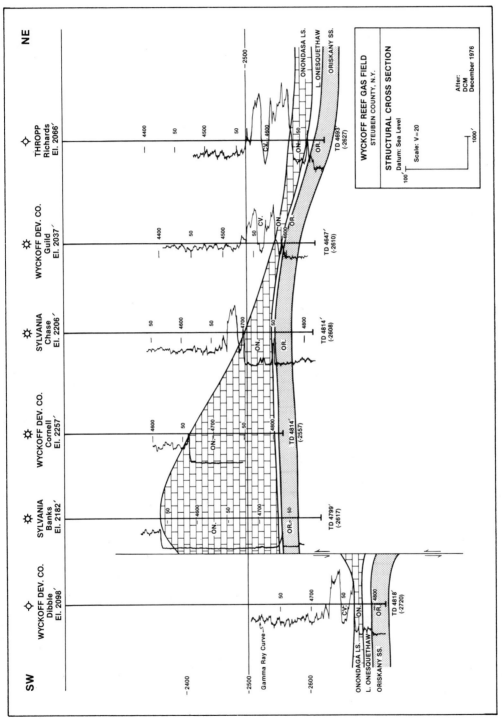

International
Human
Resources
Development
Corporation

0-934634-22-X

Randall Library – UNCW
QE501.4.C6 L3
Langstaff / Geologic cross sections
NXWW
3049002767566